Applications of Modern RF Photonics

For a complete listing of titles in the
Artech House Applied Photonics Series,
turn to the back of this book.

Applications of Modern RF Photonics

Preetpaul Singh Devgan

ARTECH HOUSE
BOSTON | LONDON
artechhouse.com

Library of Congress Cataloging-in-Publication Data
A catalog record for this book is available from the U.S. Library of Congress.

British Library Cataloguing in Publication Data
A catalogue record for this book is available from the British Library.

Cover design by John Gomes

ISBN 13: 978-1-63081-159-4

Supplemental videos for this book can be found online at:
http://us.artechhouse.com/Applications-of-Modern-RF-Photonics-P1932.aspx

685 Canton Street
Norwood, MA 02062

10 9 8 7 6 5 4 3 2 1

For Marie, Alexander, and Louise

Contents

Preface

This book is intended as an overview of the many diverse ways that RF photonics can be used to address the needs of signal-processing applications. While many previous volumes have covered the fundamental principles of RF photonics, this work focuses on the application of RF photonics to address specific needs. The book uses a demonstration architecture to highlight different areas where RF photonics can be applied. The intended audience includes senior undergraduate students studying engineering and/or physics, as well as professional researchers and engineers who are interested in using RF photonics. The book serves as a primer for RF photonic applications, making it useful for the second half of an introductory class on RF photonics. While the book presents multiple examples, it is not complete. Any omissions are not meant as a judgment on technical deficiencies, but are due simply to a limited amount of time and space.

Chapter 1 begins with a brief overview of the history of RF photonics, compares digital and analog photonics, and discusses the need for RF photonic technologies. Chapter 2 then briefly discusses the performance of the different photonic components that make up an analog delay line, highlighting where RF photonics may offer advantages. Chapter 3 covers multiple methods for improving the performance of the analog delay line; subsequently, Chapter 4 presents an overview of RF oscillators using photonic means. Chapter 5 describes methods for isolating and separating RF signals received by the demonstration architecture using photonic subsystems. Chapter 6 covers spectral analysis using RF photonic means. Chapter 7 provides an overview of RF photonics-based down conversion, analog-to digital conversion, and arbitrary waveform generation. Chapter 8 discusses the use of integrated photonic circuits. Finally, Chapter 9 provides some concluding remarks while looking at the past, present, and future of RF photonic applications.

It is my hope that this book addresses a gap in the current literature about the application of RF photonic technologies to real-world needs. This field will continue to grow, and this book should provide a basis for inspiring new designs and techniques to meet upcoming challenges.

Acknowledgments

The opinions and views expressed in this book are purely those of the author and do not reflect the official position of the Air Force Research Laboratory, the United States Air Force, or the U.S. government.

This book is a private effort, and I would like to recognize the many people who inspired and supported me in this work. Beginning with my early studies, I would like to thank Andrès, Jim, Matt, Pradeep, Mitch, Mike, Al, and all the others who challenged me and pushed me to work harder.

Following my early education, working at Lucent Technologies helped me to define my future goals and gave me the confidence to believe in my own abilities, even when I made mistakes. For everything I learned, and for all the opportunities I was given, I would like to thank Eric, Marty, Tina, Dan, Dave, Bob, Olga, Peter, Jim, Jeff, Ron, and all the others at Lucent who mentored and supported me.

During my doctoral education, I was given many opportunities to push my understanding and gain new experiences. I would like to thank Eric, Paul, Sarah, Matt, and all the others with whom I was fortunate to work. To Jay Sharping, Kobi Lasri, and Renyong Tang, thank you for all your help in the lab, for all your mentoring, and for all the fruitful discussions. To Darwin Serkland, Gordon Keeler, and Kent Geib, thank you for providing a summer spent learning new skills and reinforcing basic concepts. To William Kath and Mary Phillips, thank you for your advice and for serving on my committee. To Prem Kumar, thank you for giving me a chance and for the freedom to explore new ideas and grow as an engineer.

At the Naval Research Laboratory, I was able to fully develop a deeper understanding of the field of RF photonics. Having the ability to work with great people was always one of the best parts of the job. Thank you to Jason, Alex, Modesto, Chris, Ross, Meredith, Dave, Jim, Kristina, and the others. To Frank Bucholtz, thank you for offering me an opportunity. To John Diehl, Chris Sunderman, and Joe Singley, thank you for all the hard work, support, and positive discussions—even the heated ones. To Vince Urick, thank you for being a great technical role model and for all of your support, mentorship, patience, and commitment to the mission. To Keith Williams, thank you for giving me the freedom to pursue my interests, while always making sure they fit into the greater goals.

At the Air Force Research Laboratory, I have been fortunate to continue working in RF photonics and pursue new ideas and techniques. Having a supportive management team has helped me to expand my horizons. Thank you to Attila Szep, Steve Hary, Charles Cerny, Chris Bozada, Jesse Fanning, and Timothy Johnson. To Matt Casto, thank you for the informative discussions on analog-to-digital converters. To Dale Stevens, thank you for your unwavering support and help. To Rob Nelson, Ben Griffin, and Dean Brown, thank you for challenging my preconceived notions and helping me to see things from a different point of view. To Joe Suelzer, Tom Simpson, and Joe Haeffner, thank you for all of the technical discussions on optical injection locking and nonlinear laser dynamics. To Nick Usechak, thank you for the stimulating conversations on integrated photonics and for welcoming me into your lab without any hesitation.

Finally, none of this is possible without the support of my family. For all of their help, patience, encouragement, and love, I am forever grateful. To Alex and Louise, thank you for understanding and being the best kids I could ever ask for. To Marie, thank you for everything. I miss you.

1

Introduction to Applications of Modern RF Photonics

Radio frequency (RF) photonics is an interdisciplinary field that combines analog electronics with photonic components for the high-speed processing of RF signals. While the term RF photonics is most commonly used, this technology area is also known as microwave photonics [1–3] or analog optics [4]. Regardless of the name given, the field is focused on up-converting continuously time-varying electronic signals into the optical domain, where high-speed signal processing can occur before downconverting the signal back to the electronic domain. Signal processing is a general term that incorporates techniques to address many different applications. These applications cover a broad range, including the transmission of analog signals, the generation of low-phase noise clock signals, the isolation and frequency identification of RF signals, and many others. While these types of applications have traditionally been served by electronic means, the need to access higher frequencies and wider bandwidths has driven the search for alternatives. Because photonics can operate essentially from DC to daylight, it seems an ideal solution to address these requirements. In this introduction, a brief history covering the uses of RF photonics is presented. Next, the chapter provides an overview of the advantages of photonics versus electronics, along with a comparison between the use and performance metrics of digital and analog photonics. Finally, the chapter reviews the current needs for RF photonics.

1.1 A Brief Overview of RF Photonic History

The field of RF photonics has seen a significant growth in its exploration since the beginning of the millennium. While the field has been around since the 1970s, the adoption of RF photonic technologies has been uneven. The initial interest in RF photonics started as the use of fiber-optics for communications really began to take off. Using the same components that were developed for digital transmission, RF photonics investigated different applications. By the 1980s, the loss of optical fibers in the 1,550-nm operating window allowed them to be considered as a replacement for traditional copper wires. With the invention of the erbium-doped fiber amplifier (EDFA), however, the use of fiber-optic cables in long-haul communication networks quickly spread. While the long-haul communication networks were interested in the transmission of data over transoceanic distances, there was interest in sending analog data over optical fiber links for shorter distance applications. In the 1970s, cable television was being deployed to households in the United States. The advent of new networks that were only available through cable TV prompted a significant increase in subscribers. As this growth continued, the need to meet the demand led to the investigation of analog signals being transported over fiber-optic cables. One of the main solutions uses hybrid fiber coax (HFC) systems to deliver not only cable TV signals, but also voice [5]. With the optical fiber and amplifiers now in use, the latter part of the 1980s was devoted to improving the laser sources that could provide the needed performance for analog video transmission. This work culminated in the 1990s with analog fiber systems being readily deployed to deliver multiple TV channels to the user.

However, once digital video technology matured, the need for analog video transmission began to fade. The availability of high-definition TVs, along with other factors, helped push the network providers to deliver their programming in digital formats. While the fiber-optic networks could support either analog or digital data transmission, the stringent requirements on the components themselves were relaxed for digital transmission. Digital transmission also allowed for voice, TV, and internet traffic to be handled all on the same network without mixing analog and digital formats. This simplification led to the current connectivity that households now enjoy.

The advancements in RF photonic systems in the 1980s led to RF photonics being explored for other uses. One of the early applications was photonic beam-steering for RF phased-array antennas [6]. While fiber-optic cables' low loss and immunity to electromagnetic interference are attractive for phased arrays, their small physical size is also very important. The reduced size allows for smaller and more tightly configured antenna arrays. For a phased array that can be steered up to 90°, the spacing between the elements has to be less than half of the wavelength of the highest frequency. For example, an array that operates at 30 GHz or higher requires element spacing on the order of half a millimeter, which is much larger than the diameter of a coaxial cable. While projects explored the uses of RF photonics to feed phased arrays [7], they never realized the full potential of the technology. These results led to an investment in improving the components that made up the photonic beam-steering system. One of the key components is the external modulator. A large amount of time and money was spent to improve the performance of these devices. In addition, laser and photodetector performance was improved. These investments have now pushed the overall performance of the systems to the point that RF photonics has seen an explosion in uses.

1.2 RF Photonic Advantages

With the increased performance of photonic components developed over the last few decades, RF photonics can offer many advantages over their electronic equivalents for operating at these higher frequencies. One of the easiest comparisons to make is between the transmission lines. In RF photonics, the signal is carried over a single-mode fiber-optic cable. The single-mode optical fiber consists of a core of glass that is typically around 9 microns in diameter. The small core of glass is then surrounded by a cladding that is 125 microns in diameter. The fiber-optic cable is used to guide light generated by a laser from one point to another. The laser light has been modulated by the RF signals that are to be transported over this cable. The electronic equivalent is a coaxial cable. This cable consists of an inner conductor made of copper, which is surrounded by either a dielectric material or air. The cable then has an outer conductor. The size of the inner conductor and the amount of air or dielectric depends on the range of frequencies that the cable is designed to transport. Many different arrangements of connectors for coaxial cables exist. Table 1.1 lists the various connector types with the corresponding size of their inner conductors and the maximum frequency they support.

When fiber-optic cables were first named by Kapany [8], the losses as a function of length were quite large. However when Kao and others [9–11] were able to remove the water vapor from the glass as it was being drawn, the losses were able to approach the Rayleigh scattering limit. Figure 1.1 shows the loss of an optical fiber as a function of wavelength. The loss of optical fibers decreases as the wavelength increases. Two small peaks occur at 1,240 nm and 1,380 nm; they are both due to water absorption in the glass (silica). The minimum loss window is centered at 1,550 nm and is flat around 30 nm

Table 1.1
List of Coaxial Connectors and Their Maximum Frequency

RF Coax Connector	Maximum Frequency
SMA	18 GHz
3.5 mm	34 GHz
2.92 mm	40 GHz
2.4 mm	50 GHz
1.85 mm	70 GHz
1.0 mm	110 GHz

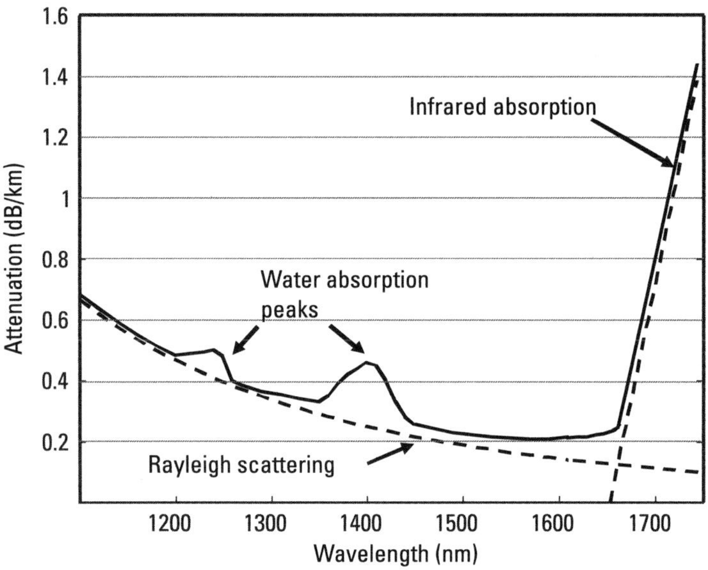

Figure 1.1 Loss of fiber-optic cable in decibels per kilometer as a function of optical wavelength from 1,100 to 1,750 nm. (After: [12].)

of that center. Above 1,620 nm, the infrared absorption of silica dominates, and the losses quickly rise. In terms of frequencies, the fiber loss is flat over approximately 3.75 THz. For a laser source operating in the low-loss wavelength window, the fiber loss is flat for any RF frequencies that are considered here.

Coaxial cables do not have a flat loss for a given length as a function of frequency. Looking at a standard coaxial cable, the losses are dominated by two effects: conductor loss and dielectric loss. The conductor loss has a square-root dependence on frequency, while the dielectric loss is linearly dependent on frequency. Figure 1.2 shows a plot of the loss of a fiber-optic cable and a coaxial cable as a function of frequency. A fiber-optic cable with a length of 1 kilometer has a loss of approximately 0.2 dB and is flat over the entire frequency range from 1 MHz to 10 GHz. A coaxial cable with a length of 10m has a very flat loss as well, which slowly rolls off to 10 dB at 10 GHz [13]. However, the loss of the same coaxial cable with a length of 100m quickly begins to grow as the frequency increases. At a frequency of 800 MHz, the loss is as high as 20 dB, or 100 times lower than the power of the signal at the input of the cable.

The bandwidth capacity of the optical fibers is also much larger than coaxial cables. Multiple lasers of different wavelengths can be multiplexed together and transmitted down a single optical fiber. Wavelength division multiplexing (WDM) has been used in multiple photonic systems to increase the data-carrying capacity of a single optical fiber. Since the loss of the fiber is flat over 30 nm (as seen in Figure 1.1), the loss each wavelength experiences is the same 0.2 dB/km. Between 40 and 80 separate wavelengths can be multiplexed, with each wavelength carrying its own data. Since these data streams are in parallel, the aggregate data rate can be quite high. The data-carrying capacity of a single fiber-optic cable is on the order of 10 Tbps over tens of kilometers, compared to a single coaxial cable carrying data on the order of 10 Gbps over tens of meters.

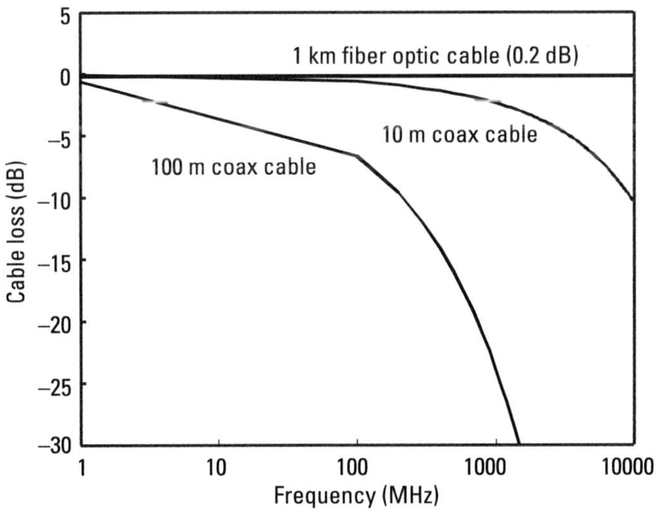

Figure 1.2 Loss of a 1-km fiber-optic cable and 10-m and 100-m coaxial cables in decibels as a function of frequency from 1 MHz to 10 GHz. (After: [12].)

While loss and bandwidth capacity are two key advantages of fiber-optic cables, there are numerous others. The weight per unit length of a coaxial cable operating out to higher frequencies can be on the order of 20–200 grams/meter, while a fiber-optic cable will weigh 1 gram/meter [13]. Thus the weight of a 1-km length of fiber-optic cable is 20–200 times less than a coaxial cable of the same length. A meter length of optical fiber costs on the order of tens of dollars while a similar length of 1-mm coaxial cable would cost on the order of a couple of thousands of dollars. Since glass expands less than copper as a function of temperature, quantified by a thermal coefficient of delay (TCD), fiber-optic cables are less susceptible to phase changes over temperature gradients than coaxial cables [14]. Optical fibers are electrically nonconductive, which makes them immune to electromagnetic interference and electrically isolates the end points of the fiber link. This can be important, as a large electrical discharge, such as a lightning strike, will not propagate down the cable and potentially destroy valuable electronics at the end of the link. Additionally, the fiber-optic cable will not spark, allowing it to be used in environments that contain highly flammable liquids, such as a fuel tank. Finally, fiber-optic cables can be easily concealed. They cannot be located using metal detectors, unlike a coaxial cable. Table 1.2 summarizes the advantages of fiber cables over coaxial cables.

1.3 Analog versus Digital Photonics

The advantages provided by fiber-optic cables are the same for both analog and digital photonic applications. The question then becomes why RF photonics did not become as widely adopted as digital photonics starting in the 1990s. The answer lies in the fact that while the fiber-optic cable and the EDFA met the performance requirements for both of the types of system, the laser, optical modulator, and photodetector did not provide the perfor-

Table 1.2
Fiber-Optic Cable's Advantages Over Coaxial Cables

	Fiber versus Coax Advantage
Loss (db/100m)	19.98 dB lower (at 800 MHz) [12]
RF bandwidth	1,000x (190 THz versus 110 GHz)
Size (inner diameter)	900x (9 microns versus 1 millimeter)
Cost ($/m)	100x ($1,000 versus $10s)
TCD (ppm/°K)	5–10x (fiber dependent) [13]
Weight (kg)	20–200x less at 1 km [12]

mance needed for analog systems. In order to understand why, the distinction between digital and analog signal processing has to be reviewed. Digital photonic signal-processing applications have been widely explored for many different commercial needs. The first and most common use is for transmitting large amounts of digital data over transoceanic distances. As opposed to analog data, digital data consists of discrete voltage levels. For binary digital data, the discrete voltage levels are denoted as ones and zeros. Typically, a one is when the voltage is greater than some threshold level, and a zero is when the voltage is below this threshold. The bit rate is given by how long in time the voltage stays a one or a zero before it changes. In Figure 1.3(a), an ideal data stream is shown. The levels of ones and zeros are clearly above and below the threshold, denoted by a horizontal dashed line. In addition, the transition between the two levels is instantaneous. Thus, the voltage stays at the intended level for the entire bit time, as denoted by the vertical dashed lines. However, a practical data stream does not look like the ideal one. Figure 1.3(b) shows a more practical data stream. In this case, the voltage levels are not always the same. While the ones and zeros are above and below the threshold line, respectively, they can often be very close. A smaller change in levels can lead to errors, where a one is identified as a zero and vice versa. Additionally, the transition between levels is not instantaneous in a practical system. A finite time is needed for the transition to occur. This reduces the time that the voltage will stay at the one or zero level. Errors occur when the measurement of the level is taken during this transition time and the wrong level is chosen.

An actual measurement of the data will yield a bit error rate (BER). Using a BER tester (BERT) will actually measure the errors of the digital data stream. By generating a pseudorandom bit stream at the input of the link, the BERT will recover the bits at the output of the link and compare them to the original bit stream. The BERT will then count the number of errors and calculate a BER. This is a measure of how many errors are found in a given unit of time. As an example, a BER of 1×10^{-10} means that one error occurs every 10^{10} bits. For a bit rate of 10 Gbps, you would on average expect to see one error every second. So the error rate is a function of the rate of the data that is being checked. However the BERT does not provide any insight on

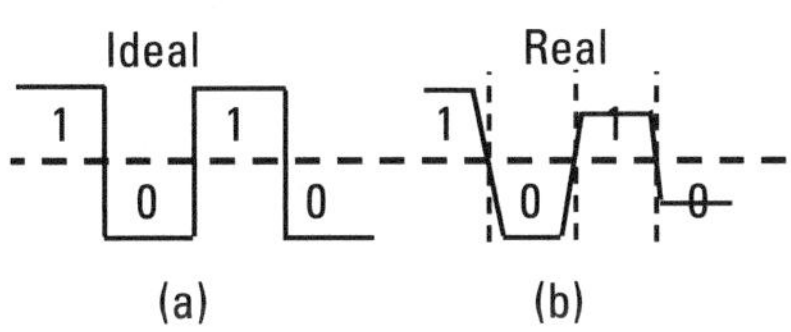

Figure 1.3 (a) Ideal binary digital data stream and (b) real binary digital data stream.

the source of the errors that occur in the link. For this, a visual representation of the data would be useful.

Using a sampling oscilloscope, an eye diagram can be measured instead. The eye diagram captures many of the important metrics for digital data by making a histogram of the measured digital data. An eye diagram gets its name from the shape of the diagram. The slope of the transitions gives the open area an oval shape, very much like the shape of a human eye [15]. Figure 1.4 shows an eye diagram for a practical data stream. The one and zero levels are not narrow lines but have a finite width to them. This width is a measure of the amplitude noise on each level. The larger the noise, the wider the level and closing of the eye diagram. The transitions between levels are also shown. The noninstantaneous rise and fall time is shown as the slope of the transitions. The width of the rise and falling is a measure of how much timing jitter, or uncertainty, there is when the transition occurs. Again, as the timing jitter increases, the width of the transitions increases and the eye begins to close. An eye mask, as shown inside the open spaces of the data stream in Figure 1.4, is a visual measure of the openness of the eye diagram. Whenever measured bits appear within the eye mask, they will lead to errors in the received data stream. Looking at the ratio of voltages for the ones and zeros level as compared to the relative widths of each level, a Q-factor can be defined. Equation (1.1) shows the *Q* factor as the difference in mean voltages between the one (μ_1) and zero (μ_0), divided by the sum of the standard deviations of the voltages of the two levels (σ_1 and σ_0).

$$Q = \frac{\mu_1 - \mu_0}{\sigma_1 + \sigma_0} \tag{1.1}$$

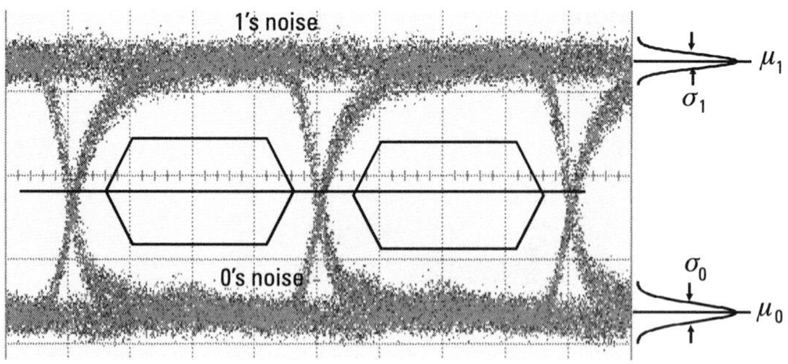

Figure 1.4 Eye diagram of a binary digital data stream. (After: [16].)

Assuming the noise has a Gaussian distribution, the Q factor can be used to estimate the BER of the data. The relation between the Q factor and BER is shown in (1.2), where the *erfc*() function is the complementary error function. While not as exact as the BERT, the visual information provided by the eye diagram is very useful for understanding the performance of the digital system.

$$\text{BER} = \frac{1}{2} erfc\left(\frac{Q}{\sqrt{2}}\right) \tag{1.2}$$

For an analog system, the key metrics are different from the digital equivalent. While a digital link is defined ultimately by the BER, the analog link performance can be distilled down to four metrics: RF gain, minimum detectable signal (MDS), compression dynamic range (CDR), and max input spur-free (MISF) power [17]. The MDS is the input power at which the output power of the system will be above the noise floor plus some margin. The formal definition of MDS appears in (1.3). The MDS is directly proportional to the RF noise figure (NF_{RF}) of the system. The MISF is the maximum input power before any output spurious signal power appears above the noise floor of the system. The definition of MISF is given in (1.4).

$$\begin{aligned}\text{MDS}(\text{dBm}) = {} & N_{\text{th}}(\text{dBm}) + \text{NF}_{\text{RF}}(\text{dB}) + 10\log_{10}(\text{BW}(\text{Hz})) \\ & + \text{Margin}(\text{dB})\end{aligned} \tag{1.3}$$

$$\text{MISF}(\text{dBm}) = \text{MDS}(\text{dBm}) + \text{SFDR}(\text{dB in1Hz}) - \text{Margin}(\text{dB}) \tag{1.4}$$

The MISF is directly proportional to the spur-free dynamic range (SFDR). As an equivalent to the eye diagram for digital systems, a plot of the analog response of an RF photonic system can be made that visually captures the metrics for these systems. Such a plot appears in Figure 1.5, which shows the relationship between the input RF power (X-axis) and output power (Y-axis) in units of decibels relative to 1 mW or log scale. The noise floor of the system is measured and displayed as a horizontal line. The power of the noise floor in relation to the thermal noise limit and the RF gain of the system is the RF noise figure. The fundamental response for a given operating frequency of the analog system is represented by a line with a slope of one. As the input power increases, the output power increases by the same amount. Simply tak-

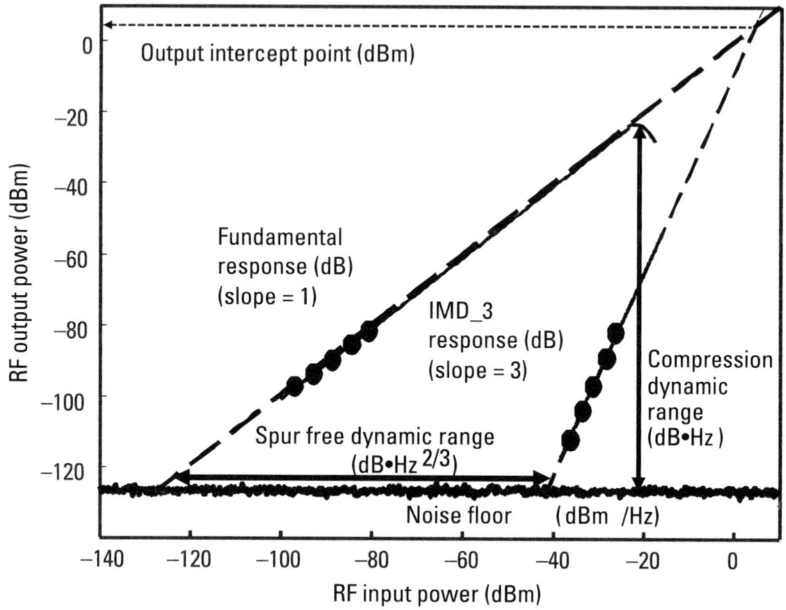

Figure 1.5 RF metrics represented by a plot of RF input power versus output power. (After: [12].)

ing the difference between the input and output powers along this line yields the RF gain.

As the input power increases, the measured output power will begin to deviate from the ideal slope of one. The deviation occurs when the system enters compression and can no longer provide the same amount of gain. When the output power is 1 dB less than the output power predicted by the slope of the line, the 1-dB compression point has been reached. The range of input powers over which the signal is above the noise floor and equal to the 1-dB compression point is the CDR. Equivalently, the range of input powers over which the signal is above the noise floor but the spurious signal, represented by the third-order intermodulation distortion (IMD_3) response line in Figure 1.5, is below the noise floor. This range of input powers is the SFDR. Finally, assuming the fundamental response did not roll over at the compression point, the line would continue until it crosses the IMD_3 line. The output power where these two lines cross is the output-referenced intercept point.

Note that digital photonic systems have a lot of advantages. Since the data is digitized, or transformed into a stream of discrete one and zero voltages, the potential exists to correct any errors that occur as the data passes over long distances. Multiple methods exist for correcting errors in transmission of digital data. One method is called forward error correction (FEC). Similar to the idea of a parity bit, FEC utilizes extra bits to create a Reed-Solomon code that can distinguish when errors occur in transmission [18]. Unlike a parity bit, the codes can actually be used to correct the errors at the end point, thus earning the name FEC. FEC requires extra overhead in order to work. For a SONET optical channel denoted as 9.953 Gbps or OC-192, the FEC-enabled data rate is actually 10.664 Gbps. The higher speed allows for the extra FEC overhead bits to be incorporated in the SONET network. FEC has been shown to take raw BERs as high as 10^{-4} and correct them to BERs of 10^{-10}, which is enough to be used at the end terminal point. The drawback is that the data has to be processed in order to correct the errors. So it is useful at the end point where the added processing time is not a hindrance.

Other methods exist to correct errors within the network that require less processing time. Regenerating the signal is one example of this type of error correction. Digital photonic networks often use 3R regeneration. 3R regenerators perform three operations that all start with "R": retime, reamplify, and reshape. Retiming corrects timing errors that can occur. It minimizes the width of the transitions in the eye diagram. Reamplification corrects errors by increasing the level of the ones, moving it further above the threshold voltage. Reshaping involves reducing the width of the one's and zero's levels. It can also be used to improve the slope of the transitions between levels. Regeneration has been a key factor in allowing digital data to be corrected before being further transmitted.

For analog transmission, the only regeneration that can be done is reamplify. The ability to retime and reshape the data does not really exist for analog transmission. This is due to the fact that digital data is known a priori to entering the network. Since the original signal is digitized into the ones and zeros, all the information is encoded. For an analog transmission system, there is no knowledge of what the signal is before it enters the network. Therefore, no retiming can be performed as there is no known fundamental frequency that the signal will follow. In fact, the analog signal may be purposefully changing frequencies over time. The analog network needs to be transparent to this effect and simply pass the signal without adding any other distortions. Due to the more stringent requirements, the lasers, optical modulators, and photodetectors used in digital applications were not able to provide the required performance. However, the photonic component performance has now improved in order to meet these requirements. Thus, RF photonics can now address the needs of current systems.

1.4 Current Needs for RF Photonics

While the component performance was improved, new applications for RF photonics were also becoming apparent. One example of the need for operating at higher frequencies is in wireless communications. Mobile telephones dominate the frequency bands from 600 MHz to 2.1 GHz in the United States. The 2.4-GHz band is also used by wireless LAN routers as well as some cordless telephones. The latest 802.11ac WiFi standard now uses frequencies at 5 GHz. Satellite communications are expanding out to 10 GHz and higher. As more and more wireless signals take up the available frequency spectrum, the only available bandwidth will be at higher frequencies. In addition, higher carrier frequencies will enable faster data rates for wireless devices in order to meet the needs of users. To meet these needs, communications will make use of frequencies above 20 GHz. As seen in Figure 1.6, the loss of a signal that is transmitted through the atmosphere is very low (< 1 dB/km) from 1 to 100 GHz, except for the rather large absorption peak around 60 GHz [19]. Clearly there is room to grow in the future. While RF photonics may not be used in the wireless transmission of data, it will be used to connect the wireless towers to the base stations and the base stations to the central station. In addition, the technology can be used to generate RF carriers at frequencies much higher than used currently, as well as to separate transmitted and received signals.

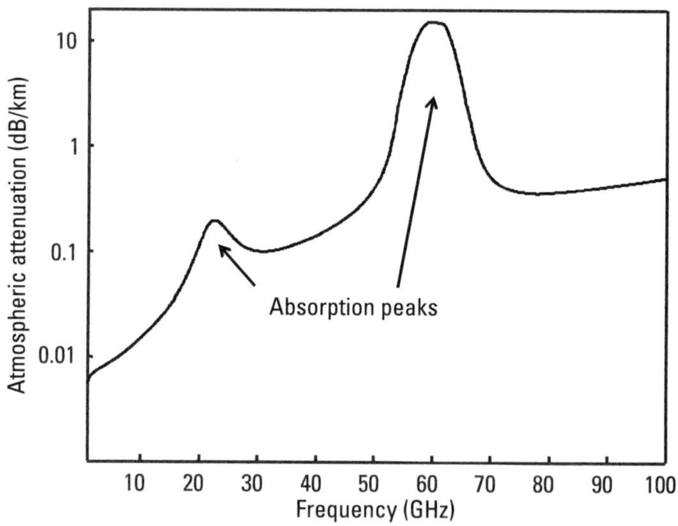

Figure 1.6 Atmospheric attenuation in decibels per kilometer as a function of frequency from 1 to 100 GHz. (After: Devgan [12].)

While the atmospheric attenuation can be low over a wide frequency range, wireless signals experience higher attenuation due to the materials used in buildings. The loss increases as the frequency of the wireless carrier increases. For a large apartment building, a hybrid method may be used to supply each apartment with access to a wireless signal. From the central station, a fiber-optic cable can be run to each apartment. The optical cable will feed an antenna in the apartment, thus creating a local wireless access point. With a proper choice of radio frequency, the signal from the access point will be attenuated enough so that it does not interfere with any neighboring apartment's signal. Thus, each apartment will act as a picocell. An RF photonic system is best suited to address this application, due to the high-frequency (HF) radio signals that are used. These specific types of links are often referred to as radio-over-fiber links [20].

RF photonic technologies have applications for needs other than data transmission. Another common application is the distribution of RF signals from one point to another. RF photonic links are currently used in radio astronomy, which utilizes large dish antennas to collect very faint radio waves from distant stars. In order to improve the signal-to-noise ratio (SNR) of the received signals, multiple antennas are used. The antennas are spread out over multiple kilometer distances from each other. In order to recover the detected signal, each antenna is connected to a central base station. Due to the large separation between antennas, RF photonic links are used to connect them together. This is just one example of antenna remoting. Other systems also use RF photonic links for antenna remoting. In some cases, the antennas can be placed in hostile environments. With coaxial cables, the antennas can only be separated from the central station by about 1,000 feet. This places the people in the central station in the same environment as the antennas. With an RF photonic link, the distance can be increased to several kilometers. One such demonstration has sent signals in the HF band over seven kilometers [21]. Other needs that have been identified include beamforming in radio astronomy [22] and true time delay [23]. All of these needs can be addressed with RF photonic solutions. Later chapters will expand on these and other applications where RF photonics provides distinct advantages over electronic solutions.

1.5 Conclusions

Analog systems do offer advantages over digital systems in some cases. When the frequency of the signal is higher than can be digitized with the required fidelity, an analog system must be used. Analog systems are also often more robust than digital systems. In a measure of the reliability of links for a radio astronomy system, the analog links had a fourfold increase in the mean time between failures when compared to digital links [24]. In the digital link, the analog-to-digital converter (ADC) is placed right at the antenna, exposing it to the elements. ADCs often failed due to environmental effects, such as lightning. Analog links have less complex front ends, leading to a less failures. Clearly, analog systems are useful in certain applications. RF photonic systems that offer advantages over other solutions are explored in further details in the following chapters.

The remainder of the book focuses on the various applications of RF photonics. These include examples of analog delay lines, oscillators, signal isolation, and filtering. The various applications are reviewed in the context of an analog architecture that requires these various functions. In addition, we will develop a demonstration architecture over the course of the book. The different RF photonic signal-processing applications that can be accomplished within the architecture are highlighted in individual chapters. The initial architecture will consist of an antenna that receives an RF spectrum of signals and passes those signals through to a final output. Subsequently, we will explore additional functionality to add to the architecture. While not all applications will be covered by the architecture, it will provide a concrete example of how RF photonics can address different applications.

References

[1] Seeds, A. J., and K. J. Williams, "Microwave Photonics," *Journal of Lightwave Technology,* Vol. 24, No. 12, 2006, pp. 4628–4641.

[2] Capmany, J., and D. Novak, "Microwave Photonics Combines Two Worlds," *Nature Photonics,* Vol. 1, No. 6, 2007, pp. 319–330.

[3] Vilcot, A., B. Cabon, and J. Chazelas (eds.), *Microwave Photonics: From Components to Applications and Systems,* Boston, MA: Kluwer Academic Publishers, 2003.

[4] Cox III, C., *Analog Optical Links,* Cambridge, England: Cambridge University Press, 2004.

[5] Kaminow, I. P., and T. Koch (eds.), *Optical Fiber Telecommunications IIIA,* Cambridge, MA: Academic Press, 1997.

[6] Williamson, R., and R. Esman, "RF Photonics," *Journal of Lightwave Technology,* Vol. 26, No. 9, 2008, pp. 1145–1153.

[7] Yao, J., "A Tutorial on Microwave Photonics," *IEEE Photonics Society Newsletter,* June 2012, pp. 5–12.

[8] Kapany, N. S., *Fiber-optics Principle and Applications,* Cambridge, MA: Academic Press, 1967.

[9] Kao, K. C., and T. W. Davies, "Spectrophotometric Studies of Ultra Low Loss Optical Glasses-I: Single Beam Method," *Journal of Physics E: Scientific Instruments,* Vol.1, No. 11, 1968. pp. 1063–1068.

[10] Jones, M. W., and K. C. Kao, "Spectrophotometric Studies of Ultra Low Loss Optical Glasses-II: Double Beam Method," *Journal of Physics E: Scientific Instruments,* Vol. 2, No. 4, 1969, pp. 331–335.

[11] Wright, C. R., and K. C. Kao, "Spectrophotometric Studies of Ultra Low Loss Optical Glasses-III: Ellipsometric Determination of Surface Reflectances," *Journal of Physics E: Scientific Instruments,* Vol. 2, No. 7, 1969, pp. 579–583.

[12] Devgan, P., "Review of Radio Frequency Photonics Basics," AFRL Report, AFRL-RY-WP-TR-2017-0157, 2017.

[13] MIL-C-17 Attenuation loss and power handling tables.

[14] Dreher, J., *Phase Stability of ATA Fiber-optic Cables,* Seti Institute, ATA Memo 55, March 2003.

[15] Agrawal, G. P., *Fiber-optic Communication Systems,* Hoboken, NJ: John Wiley and Sons, 1997.

[16] Devgan, P., "High-speed Signal Processing Using Nonlinear Fibers and Optoelectronic Devices," Ph.D. dissertation, 2005.

[17] Pozar, D. M., *Microwave Engineering,* Hoboken, NJ: John Wiley and Sons, 2009.

[18] Chow, M-C., *Understanding SONET/SDH Standards and Applications,* Holmdel, NJ: Andan Publisher, 1996.

[19] ITU Recommendation, *Attenuation by Atmospheric Gases,* ITU-R P.676-10, Aug. 2013.

[20] Berceli, T., and P. Herczfeld, "Microwave Photonics—A Historical Perspective," *Microwave Theory and Techniques,* Vol. 58, No. 11, 2010, pp. 2992–3000.

[21] Urick, V. J., et al., "Field Test on the Feasibility of Remoting HF Antennas with Fiber-optics," *NRL Memorandum Report,* NRL/MR/5652-08-9137, 2008.

[22] van Ardenne, A., et al., "A Dense Aperture Array with Photonic Beamforming For Radio Astronomy," *International Conference on Electromagnetics in Advanced Applications (ICEAA),* 2014.

[23] Zach, S., and L. Singer, "RF Photonics—Why Should Defense Take Notice?," *Convention of Electrical and Electronics Engineers in Israel,* 2006.

[24] Montebugnoli, S., et al., "Large Antenna Array Remoting Using Radio-Over-Fiber Techniques for Radio Astronomical Application," *Microwave and Optical Technology Letters,* Vol. 46, No. 1, 2005, pp. 48–54.

2

Analog Delay Lines

One of the first applications where RF photonics has proved to offer advantages is a simple RF link. The link is one of the easiest applications to understand when first discussing RF photonic technologies. The RF photonic link simply needs to transport a spectrum of signals from one point to another. One example of an RF photonic link would be connecting an antenna to a signal-processing system, as shown in Figure 2.1. A laser provides an optical carrier to a photonic modulator. The RF signal is then upconverted onto the optical carrier via the photonic modulator and then sent down an optical fiber cable. At the end of the link, a photodetector downconverts the RF signal to the electrical domain. For this application, the photonic link is designed to deliver the RF signal at the end point with as little distortion as possible. When compared to a coaxial cable link, the RF photonic link may not always be able to match the nonlinear performance. As the coaxial cable is a passive transmission line, it adds no nonlinearities to the signals. However, the RF photonic link is an active link. Since the link has an upconversion operation followed by a downconversion operation, there are nonlinear distortions added. Often, the choice between a coaxial cable link and an RF photonic link comes down to needs. RF photonic links are best suited for either long lengths or ultrawide frequency operations. For short lengths (< 100m) and low frequencies (< 1 GHz), coaxial cables are more than capable of handling the needs of most RF transmission links. However, when the length increases or the frequency exceeds these bounds, the RF photonic link is often the preferred choice. While the coaxial links do not add nonlinearities, they do

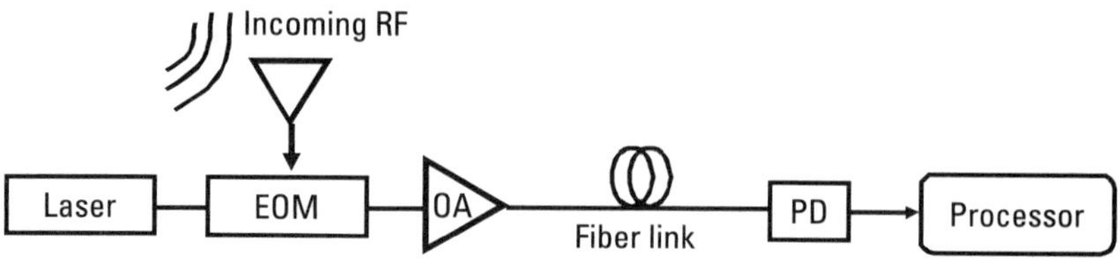

Figure 2.1 Example of an RF photonic link.

suffer from very high losses as either the length or the frequency increases, as seen in Figure 1.2. One specific application where RF photonic links have an advantage over coaxial links is analog delay lines.

An analog delay line has many applications, from emulating long-distance transmission to acting as an analog memory unit. Analog delay lines meet the thresholds for using RF photonic technologies, as they often have large frequency bandwidth requirements and require long lengths. In order to quantify the performance of an analog delay line, the RF metrics need to be addressed. While the key RF metrics can be distilled down to the MDS and the MISF, these are based off the RF gain, noise figure (NF), and SFDR. Phase noise and CDR are two other important RF metrics. To design an RF photonic delay line, the predicted RF metrics must be calculated. RF photonic links can be broken down into their constituent components, namely the laser, modulator, optical amplifier, optical fiber, and photodetector. As these components drive the overall RF performance of the delay line, a translation between photonic performance and RF performance needs to be defined. This chapter defines the various RF metrics, both inherently and as a function of photonic component performance.

2.1 Different Examples of Analog Delay Lines Using RF Photonics

As discussed previously, analog delay lines require different performance metrics in order to be usable for their intended applications. A few examples of the possible uses of analog delay lines can highlight the metrics that need to be accounted for when designing RF photonic delay lines. In one such example, the RF photonic delay line can be used as an analog buffer, and this is how we will use the analog delay line in our demonstration architecture. As seen in Figure 2.2, a spectrum of RF signals can be captured from an antenna and split into two different paths. The first path delivers the signals to the processing system, which will process the signals. One example of this type of processor is a thresholder that identifies signals above a certain power level. Any signal that meets this requirement can then be digitized. However,

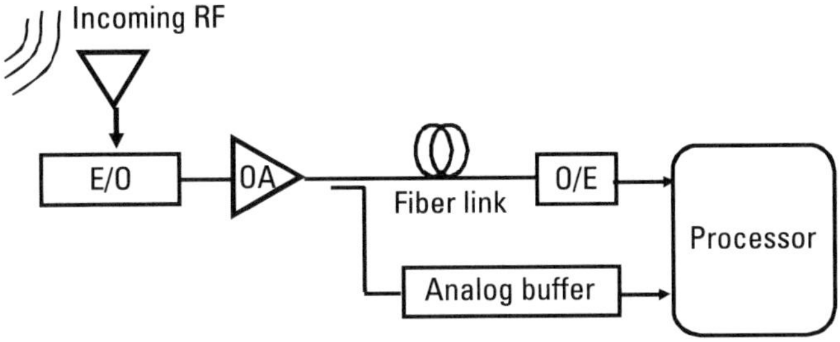

Figure 2.2 Demonstration architecture incorporating an analog buffer.

the original signal may not be able to be digitized after the thresholder as it may be significantly distorted or attenuated by the thresholding process. If a copy of the signal was kept, then the unmodified signal can be digitized. By splitting the spectrum of signals, an original copy is preserved in the analog buffer. The analog buffer just needs to hold the spectrum long enough for the processing system to identify the chosen signal and then set up the digitizer to record the signal. The required processing time can be on the order of 100 μs. An analog buffer would have to hold the spectrum of signals for at least that length of time. To achieve this buffering time, an analog delay line would need to be constructed with a delay length of approximately 20 km. A coaxial delay line would not be able to meet this requirement as the loss would be too large to properly preserve the original signals. Fortunately, an RF photonic-based delay line would be able to meet these requirements. In this case, the key RF metric of interest is the NF of the delay line. Since the signal power can be conditioned (amplified or attenuated as necessary) at the input of the delay line, the signals can be prepared to stay below the MISF and above the MDS. Even if the input power is above the MISF, any spurious signals that are generated will be ignored as they would not match the frequency of the signal to be digitized. However, if the delay line has too high of a NF, the SNR will be sufficiently degraded. The digitizer will then be unable to capture any useful information. So, in the design of this delay line, the RF NF can be optimized at the expense of the SFDR.

In another case, the RF photonic delay line can be used as a range emulator. In this case, a user wishes to transmit a single or limited set of signals over a long distance (> 10 km). Multiple reasons exist for not actually transmitting the signal over that long of a distance, including not having access to an open area of that distance and the complexity and cost of setting up a transmitting and receiving station over a significantly long distance. Instead, an analog delay line can emulate the long distance in a small form factor, allowing the signal fidelity after transmission to be compared with the original in the same location. For this application, the delay line needs to add as little distortion as possible to the signals. Since the signal is known going into the delay line, the input SNR can be properly set at the input so that after the delay line the SNR is sufficient for the signal performance to be determined. Therefore, the NF of the delay line is important, but not as important as the nonlinear distortion. In this case, the delay line is optimized for SFDR and the amount of added phase noise is reduced.

2.2 Definitions of RF Metrics

The five main RF metrics that define most RF photonic systems are the RF gain, RF NF, SFDR, CDR, and the phase noise. With these metrics, the performance of the RF photonic system can be calculated and used to determine if it is appropriate for the application. The first metric that all RF applications need to know is the RF gain of the system. The RF gain is a measure of the RF power at the output of the system relative to the input RF power to the system. For this discussion, the RF gain will be defined in logarithmic units of decibels. Note an RF gain of 0 dB is defined as a unity gain, and a negative RF gain means that the system has loss. The mathematical expression of the RF gain appears in (2.1a). For a complex RF system that contains multiple RF components, the total RF gain is simply a sum of the various gains of the individual components, also shown in (2.1b).

$$G_{rf}(dB) = P_{rf,out}(dBm) - P_{rf,in}(dBm) \tag{2.1a}$$

$$G_{tot}(dB) = \sum_{i=1}^{n} G_i(dB) \tag{2.1b}$$

The RF NF is measured as the change in the SNR at the input of the system in relation to the SNR at the output. Note that the noise at the input of the system must be thermally limited. If this is not the case, the definition of NF is no longer true. The mathematical definition of RF NF is presented in (2.2a), where k_B is Boltzmann's constant, T is the temperature (typically defined at 290K), and N_{out} is the measured output noise level. Note that the bandwidth is assumed to be 1 Hz in this definition. The NF of a composite system can be determined by using the NF of the individual components that make up the system. The Friis NF cascade equation is shown in (2.2b) [1]. Looking at (2.2b), the overall system NF is often determined by the first component's NF, as long as the gain of the first component is enough to compensate the NF of the following components. In most RF systems, a low noise amplifier is the first component in the system. The low NF, combined with high RF gain, sets the overall NF performance of the system.

$$NF_{\mathrm{rf}} \equiv \frac{SNR_{\mathrm{in,th}}}{SNR_{\mathrm{out}}} = \frac{S_{\mathrm{in}} \cdot N_{\mathrm{out}}}{k_{\mathrm{B}}T \cdot S_{\mathrm{out}}} = \frac{N_{\mathrm{out}}}{G_{\mathrm{rf}} k_{\mathrm{B}} T} \tag{2.2a}$$

$$NF = NF_1 + \sum_{\substack{i=2 \\ m \geq 2}}^{m} \left(\frac{NF_i - 1}{\prod_{p=1}^{i-1} G_p} \right) \tag{2.2b}$$

The RF SFDR is a measure of the input powers over which the original input signal is above the system noise floor while no spurious signals appear. The spurious signal is a result of nonlinearities in the components of the system, which leads to either harmonic generation from a single RF input or intermodulation distortions from the mixing of multiple frequency components. A plot of the different spurious signals that are generated from a nonlinear RF system is shown in Figure 2.3. In this case, two different single, CW frequency tones are input to the RF system. At the output, the original two signals are observed, as are several other spurious signals. For each of the two tones, the harmonic spurious signals appear. The second harmonic appears at twice the original frequencies, followed by the third, and so on. In addition, since there were two original signals, the intermodulation spurious signals appear as well. The second-order intermodulation spurious signals appear at the sum and difference of the original two signals. The third-order intermodulation spurious signals appear at the difference of double one signal's frequency

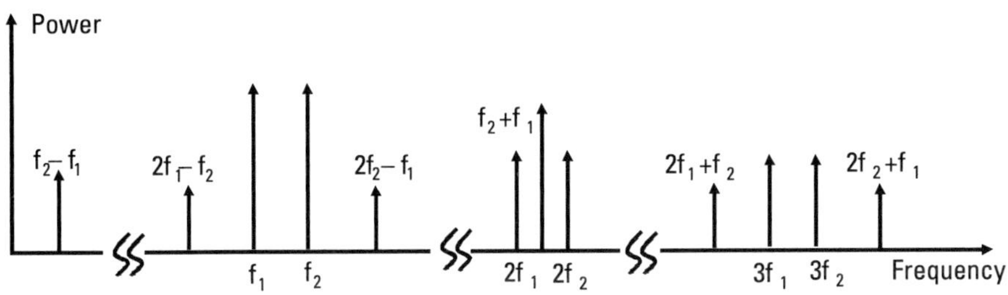

Figure 2.3 Diagram of the various nonlinear spurious signals for a two-tone input.

minus the other signal's frequency. Often RF systems are limited to a single octave of bandwidth in order to filter out the harmonic and second-order intermodulation spurious signals. However, the third-order intermodulation spurious signals appear within an octave bandwidth and cannot be filtered out. These types of RF systems are referred to as third-order-limited. One of the advantages of an RF photonic system is that they can be third-order-limited while operating over greater than an octave of bandwidth. The reason why this occurs will be discussed later in Section 2.3.

The SFDR can be calculated from two different data points, the NF of the system and the output-referenced intercept point (OIP). The OIP is measured by plotting the output power of the system as a function of the input power, as seen in Figure 2.4. For the two-tone input, the input power of each tone is increased by the same amount, and the power of each is kept equal. The plot of the two-tone output, known as the fundamental response, is a measure of the gain or loss of the RF system. The spurious signals of interest are also plotted. In the case of Figure 2.4, the second- and third-order harmonics as well as the intermodulation distortions are plotted. For the fundamental response, the line will ideally have a slope of one, while the second- and third-order spurious signal responses will have slopes of two and three, respectively, on a logarithmic scale. The OIP is the output power where the extrapolated fundamental response line crosses that of the spurious signal response. Note that the second harmonic $OIP2_H$ is 6 dB higher than the second-order intermodulation $OIP2_{IMD}$, while the third harmonic

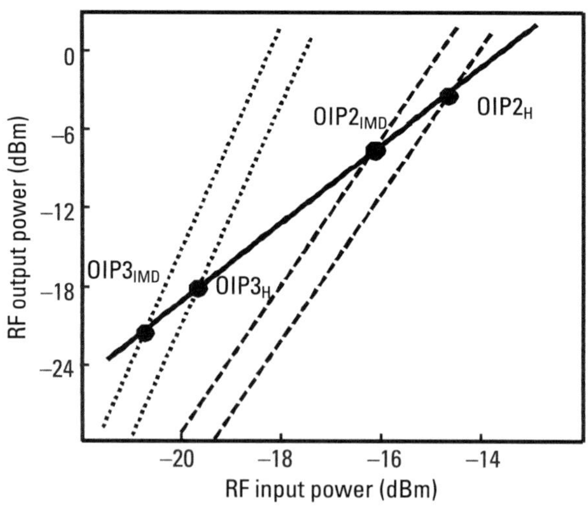

Figure 2.4 Second- and third-order nonlinear intercept points for intermodulation and harmonic distortion.

$OIP3_H$ is 4.8 dB higher than the third-order intermodulation $OIP3_{IMD}$ [2]. Once the OIP is determined, the SFDR is calculated using the output noise. Just like the NF, an overall OIP can be calculated using the OIP of the individual components that make up the RF system. The OIP cascade formula is presented in (2.3b). From (2.3b), the worst case OIP of the overall system is often set by the last element in the system, which is the opposite of the cascade result for the RF NF.

$$\mathrm{SFDR}\left(\mathrm{Hz}^{(n-1)/n}\right)=\left(\frac{OIPn}{N_{\mathrm{out}}}\right)^{(n-1)/n} \tag{2.3a}$$

$$OIP2=\left[\sum_{\substack{i=1\\ m\geq 2}}^{m-1}\left(OIP2_{i}\prod_{p=i+1}^{m}G_{p}\right)^{-1/2}+OIP2_{m}^{-1/2}\right]^{-2}$$

$$OIP3=\left[\sum_{\substack{i=1\\ m\geq 2}}^{m-1}\left(OIP3_{i}\prod_{p=i+1}^{m}G_{p}\right)^{-1}+OIP3_{m}^{-1}\right]^{-1} \tag{2.3b}$$

The RF CDR is the range of input powers over which the output signal is above the noise floor of the system and the output power follows the linear gain of the system. An example of the CDR appears in Figure 2.5; the solid line is the measured output power while the dashed black line is the ideal response line with a slope of one. The signal power is above the noise floor and follows the ideal response until it begins to roll off. When the measured output power is 1 dB below the ideal response, the input power is referred to as the 1-dB compression point. The CDR is then calculated using (2.4). In most RF photonic systems, the CDR is often much larger than the SFDR, due to the nonlinear response of the upconversion and downconversion process that occurs in these systems [3].

$$\mathrm{CDR}(\mathrm{dB}\cdot\mathrm{Hz})=P_{1\mathrm{dB}}(\mathrm{dBm})+1-N_{\mathrm{out}}(\mathrm{dBm/Hz}) \tag{2.4}$$

Finally the added phase noise is an important metric for analog delay lines. Specifically, analog delay lines that are used to emulate the transmission of a signal over a long length are often concerned with the added phase noise to the signal. Phase noise is used as a measure of how spectrally pure a

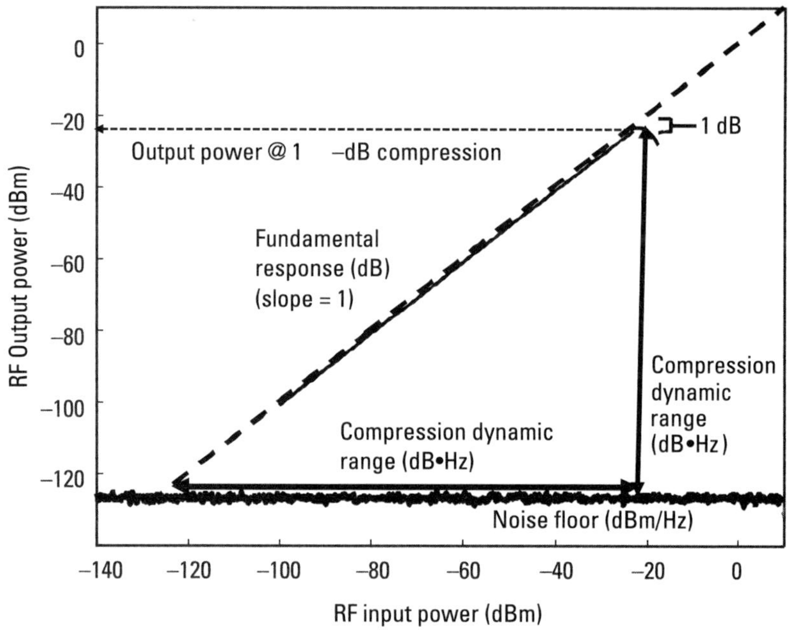

Figure 2.5 One-dB CDR. (After: [4].)

single RF tone from an oscillator is. While an ideal oscillator will generate a perfect RF tone at a given frequency, a practical RF tone will have some noise bandwidth. The phase noise of the tone determines the power of the noise relative to the power of the carrier as a function of the frequency offset from the center frequency. Phase noise is measured by taking the single sideband power spectral density of the signal's phase fluctuations [5]. The larger the phase noise, the more the RF signal will deviate from its fundamental frequency. The phase noise is also used to calculate the timing jitter of the RF signal. Phase noise, which is often used to measure the quality of an oscillator, is covered in greater detail in Chapter 4.

2.3 Different Architectures of RF Photonic Delay Lines

In order to express the important RF metrics of an analog delay line in terms of the RF photonic components that make up the delay line, the architecture has to be determined. When determining the type of architecture to use, the most important decision to be made is the type of modulation that is used. Multiple ways exist to upconvert the RF spectrum of signals onto an optical carrier. All RF photonic links require at least a laser, an optical fiber, and a photodetector. A choice needs to be made between external and direct modulation. Direct modulation acts on the laser itself, while external modulation uses a separate modulator after the laser to upconvert the signal on to the optical carrier. Direct modulation has been investigated by many for use in RF photonic links, including delay lines. Figure 2.6, which shows a directly modulated link along with a plot of the laser power as a function of drive current, illustrates how the direct modulation occurs. Once above the threshold current for lasing to occur, the optical power increases as the drive current increases. The relation between the drive current and the optical power is known as the slope efficiency of the laser. This relation appears in (2.5). Picking a bias point where the laser is above threshold, the current can be modulated by an analog signal, allowing it to be upconverted onto the power of the optical carrier. One of the advantages of direct modulation is how simple it is to implement. Unfortunately, though, direct modulation has a serious disadvantage. The RF gain from a directly modulated link does not meet most analog delay line applications. The RF gain for a directly modulated link is directly proportional to the square of the slope efficiency, s_L, of the laser ($G_{rf} \propto s_L^2$) as seen in (2.6), where K is a constant related to the link loss and the responsivity of the photodetector and Z_{in} and Z_{out} are the input and output impedances, respectively. K is always less than one. Equations (2.7) and (2.8) show the second and third harmonic OIP for the directly modulated link. The constants b and c are due to the nonlinear slope of the bias

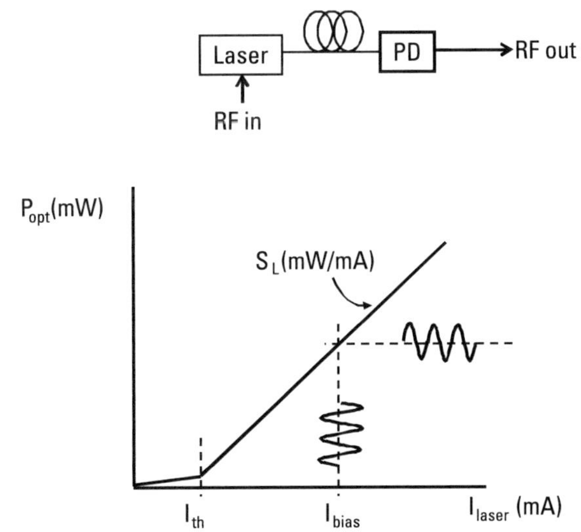

Figure 2.6 Directly modulated link architecture with laser output power as a function of bias current. (After: [4].)

curve and have to be measured for each different laser. Since a practical laser will never be 100% efficient, the slope efficiency will always be less than one. From the above relation, the RF gain will always be negative in the decibel scale if the slope efficiency is less than one. Therefore, the RF link will always introduce loss to the signal. In order to overcome this limitation, the link will require either optical or electronic amplification. Unfortunately, an amplifier will add additional noise, degrading the SNR performance of the link. Other disadvantages of direct modulation include limited frequency response and added frequency chirp to the optical carrier that can degrade the RF performance after traveling through a dispersive fiber medium [6].

$$s_L \propto \frac{I_{dc}}{I_{bias}} \tag{2.5}$$

$$G_{\mathrm{rf}} = K s_L^2 \frac{Z_{\mathrm{out}}}{Z_{in}} \tag{2.6}$$

$$OIP_{2H} = 2\frac{K}{b^2} s_L^4 Z_{\mathrm{out}} \tag{2.7}$$

$$OIP_{3H} = 2\frac{K}{c} s_L^3 Z_{\mathrm{out}} \tag{2.8}$$

Having eliminated direct modulation as a choice for the analog delay line architecture, external modulation is now left. In external modulation, multiple solutions can be used. Two main types have been investigated: phase and intensity. Phase modulators encode the analog signal onto the phase of the optical carrier, while keeping the intensity of the light constant. Phase-modulated links, as seen in Figure 2.7, have been investigated for multiple applications [7]. Phase modulation offers several advantages, including removing the need for a DC bias at the modulator as well as having lower optical loss when compared to other intensity modulators. In fact, the phase-modulated RF link can provide more RF gain at the output when compared to an intensity-modulated link [7]. However, the phase-modulated link comes with a couple of drawbacks. First, the phase modulation has to be

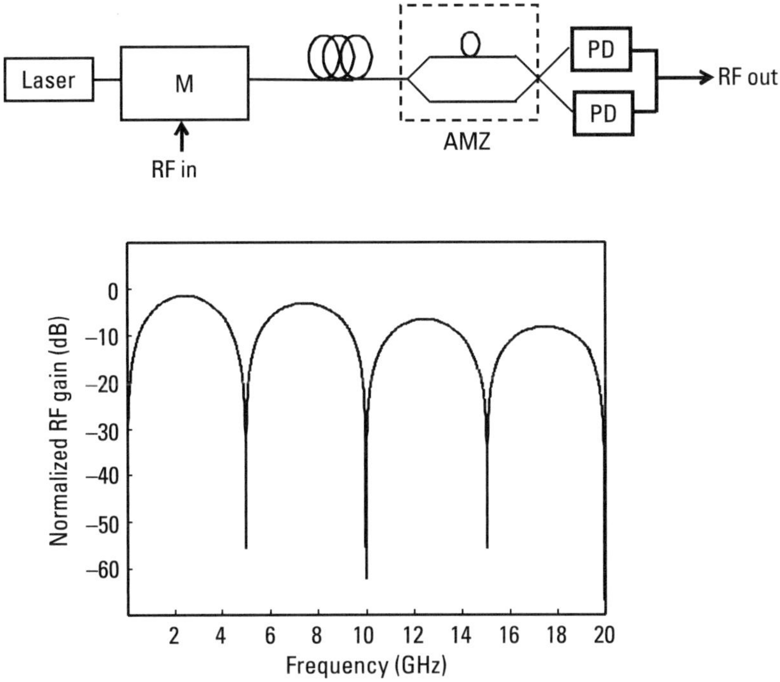

Figure 2.7 Phase-modulated link architecture with RF response of asymmetric interferometer detection with a delay of 200 psec. (After: [4].)

converted to intensity before the photodetector in order to recover the RF signal. While multiple conversion methods exist [8, 9], a common one is the use of an asymmetric interferometer for demodulation. The asymmetry in the interferometer used for demodulation places a bandwidth limitation on the link. This effect is shown in Figure 2.7. When the frequency of the RF signal is near the peak of the response, the RF gain of the link is maximized. However, as the frequency changes, the RF gain drops. This is a function of the asymmetry of the interferometer and has a response that is proportional to $\sin^2(\tau\Omega/2)$, where τ is the time difference between the two arms of the asymmetric interferometer arms (τ is 200 psec in Figure 2.7) and the Ω is the angular frequency of the signal. This response places a limit on the frequencies that can be properly sent down an RF link using phase modulation. Along with the bandwidth limitation, the phase noise of the laser will also play a role in the performance of the RF link. Like RF phase noise, the optical phase noise of a laser is often a measure of how far the light from the laser is from ideal monochromatic light. Since the laser light is passed through an asymmetric interferometer, any phase noise on the laser will be converted to intensity noise. This added noise adversely affects the RF noise performance.

The added noise is shown in Figure 2.8 [10, 11]. Both of these drawbacks limit the use of phase modulation in RF links.

Two main choices exist for intensity modulation. The first is the electroabsorption modulator (EAM), which directly affects the intensity of the light as a function of the RF signal that is applied. The EAM has previously been investigated for uses in RF photonic links [12]. The EAM link, along with a plot of the output power of an EAM as a function of applied voltage, appears in Figure 2.9. The EAM has a transfer function as shown in (2.9), where P_o is the maximum output power of the EAM, V is the applied voltage to the EAM, V_o is the voltage when the output power is 1/e of the maximum output power, and α is a constant of the EAM, with typical values between 1–2 [12]. The transfer function demonstrates the nonlinear relationship between optical power and voltage. From Figure 2.9, the slope efficiency can be greater than one, overcoming the deficiency of the direct modulation of the laser. Assuming $\alpha = 1$, a simplified version of the RF gain with an EAM link is shown in (2.10), where I_{dc} is the DC photocurrent generated at the photodetector from the incoming optical carrier, and Z_{in} and Z_{out} are the input and output impedances, respectively. The slope of the EAM is highly nonlinear, allowing for very strong spurious tones to be created. The OIP of the second and third harmonics are shown in (2.11) and (2.12), respectively. One drawback is that the EAM has a strong second harmonic response, which often limits the use of this component to single-octave operation. Other drawbacks to using the EAM are the high insertion loss along with the low optical power handling. Most EAMs can only accept a few tens of milliwatts of optical

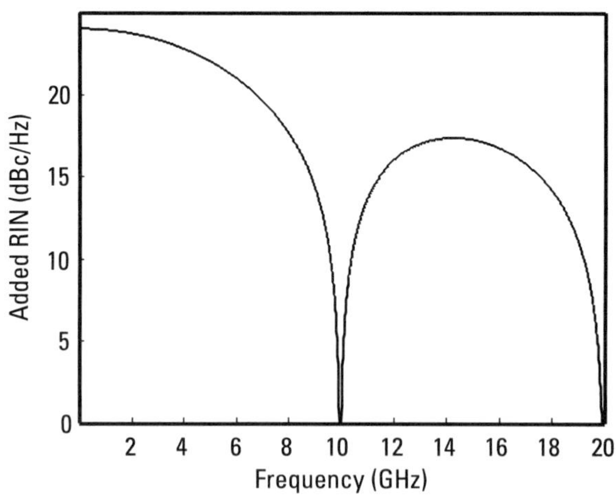

Figure 2.8 The added RIN from a 100-kHz line width after an asymmetric interferometer having a differential delay of 100 ps. (After: [4].)

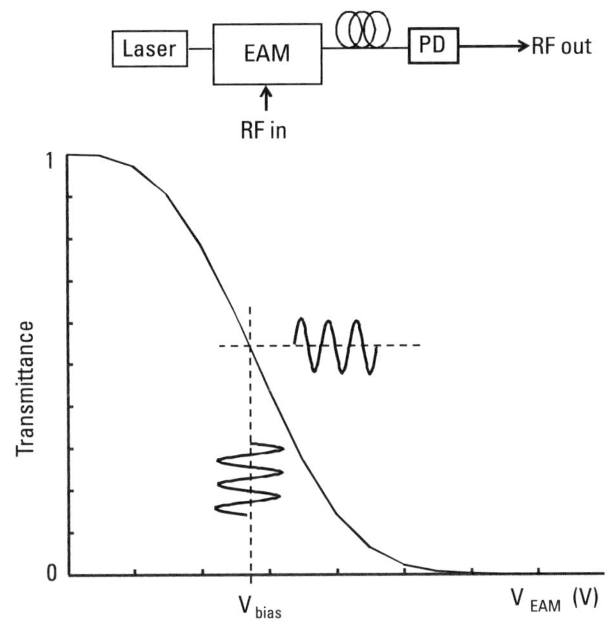

Figure 2.9 EAM link with transfer function of the EAM. (After: [4].)

power before they are permanently damaged. While recent advances have increased the optical power handling up to 100 milliwatts, the insertion loss is still large. Finally, the EAM requires temperature control as the semiconductor material is very temperature-dependent.

$$P_{out} = P_o e^{-\left(V/V_o\right)^{\alpha}} \tag{2.9}$$

$$G_{rf} = \frac{I_{DC}^2}{V_o^2} Z_{out} Z_{in} \tag{2.10}$$

$$OIP_{2H} = \frac{1}{4} I_{dc}^2 Z_{out} \tag{2.11}$$

$$OIP_{3H} = 12 I_{dc}^2 Z_{out} \tag{2.12}$$

The other commonly used external intensity modulator is the Mach Zehnder interferometer-based modulator (MZM). While other modulators also exist, they often have the same limitations as the modulators listed above.

For these reasons, the lithium niobate-based ($LiNbO_3$) MZM is the preferred solution for externally modulated RF photonic applications. The $LiNbO_3$ material requires no external temperature control and can handle well over 100 milliwatts of optical power. The insertion loss can be made as small as 3 dB, and the modulator can be driven with up to a watt of RF power before failing. Finally, the MZM configuration can be operated at a bias point where the RF even-order nonlinearities are nulled. For all of these advantages, the MZM is the chosen modulator for most RF applications.

The sinusoidal transfer function of the MZM, along with the link architecture, is shown in Figure 2.10. The MZM operates by splitting the light between two parallel paths. On each path, the applied RF signal acts through the electro-optic effect in $LiNbO_3$ to either retard or advance the light. This results in a complementary phase modulation of the light in each arm. The coupler at the end of each path then recombines the lights, allowing them to interfere with each other. The resulting output from the MZM is an intensity modulation of the light from the laser. From a traditional RF point of view, the MZM is simply a mixer, yet one that operates at 192 THz. The optical carrier from the laser mixes with the RF signal to upconvert the signal into the optical domain. In the frequency domain, the result is double-sideband modulation of the optical carrier, as seen in Figure 2.10. The sidebands are symmetrically separated from the optical carrier by the frequency of the RF signal. After transmission of the modulated light, the photodetector simply downconverts the RF signal from the optical domain back to its original

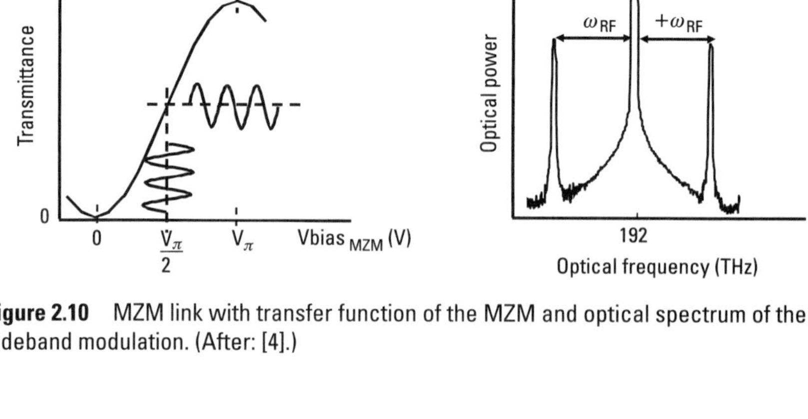

Figure 2.10 MZM link with transfer function of the MZM and optical spectrum of the dual sideband modulation. (After: [4].)

frequency. Since the optical carrier arrives at the photodetector along with the sidebands, no external local oscillator is required to recover the RF signal. This type of RF photonic link is referred to as an intensity-modulated direct detection (IMDD) link.

One of the key advantages of the IMDD link using an MZM is the linear performance. When the DC bias point that provides a phase shift of $\pi/2$ is chosen, the even order terms of the nonlinear transfer function becomes zero. This operating point is often referred to as the quadrature bias point. One of the drawbacks of using an MZM is keeping the modulator biased at quadrature. Often the bias point will drift and requires an external bias control circuit to keep the MZM at the quadrature bias point. Two popular methods exist for keeping the MZM at the quadrature bias point: optical power monitoring or RF second harmonic power monitoring.

The optical power monitoring works for a dual-output MZM by tapping off a small amount of the light from each of the optical outputs. The ratio between the powers is then used in a feedback loop to keep them at the same ratio. Ideally, the power in both arms would be equal, but for practical MZMs that do not have an infinite extinction ratio, the ratio has to be determined for each MZM to remain at quadrature bias. For an MZM that does not have two optical outputs, the optical power ratio is measured between the input of the MZM and the output. Either method requires measuring the losses of the optical taps as well as the loss of the MZM and then setting the appropriate ratio for the feedback loop to maintain the quadrature operating point. This type of bias controller is often referred to as ditherless control.

The RF second harmonic power monitoring technique works by introducing a known RF frequency into the MZM. The bias controller then taps off a small amount of the light from the optical output and measures the power of the second harmonic of the RF frequency that was injected into the MZM. The feedback loop acts to minimize the second harmonic power, which ideally occurs at the quadrature bias point. This type of bias controller is referred to as dither-based control, since a frequency dither or tone is added to the MZM. The dither bias controller can automatically find the quadrature bias point without any manual tuning. The disadvantage is that the addition of an RF tone can potentially corrupt the RF signals that are being stored in the analog delay line. The choice of bias controller depends on the application. The RF tone can also be used to improve the RF performance of the photodetection process, as discussed in Chapter 3.

The resulting RF metrics for this type of link assume that the quadrature bias point is selected. As it turns out the RF gain is also maximized when the link is operated at the quadrature bias point. The resulting RF metrics are shown (2.13a)–(2.13c), which assume that shot noise is the dominant noise term. In these equations, e is the electron charge, V_π is the modulation efficiency of the MZM, I_{dc} is the DC photocurrent generated at the photodetector from the incoming optical carrier, and Z_{in} and Z_{out} are the input and output impedances, respectively. The equations offer a translation between RF performance and the photonic metrics of the link. From a cursory glance, the main photonic metrics that improve the RF performance are the V_π and the I_{dc}. As an example, the RF gain as a function of DC photocurrent at different V_π levels is plotted in Figure 2.11. Clearly, the higher the DC photocurrent, the higher the RF gain, while the lower the V_π the higher the RF gain as well. The MZM not only plays a significant role in determining the RF gain, but also in the SFDR. From Figure 2.10(a), as the V_π decreases, the linear portion of the sinusoidal transfer function shrinks. This in turn reduces the SFDR. The tradeoff in RF gain, NF, and SFDR as a function of V_π has been discussed in other works [13]. Knowing that the V_π can only be reduced so far, the DC photocurrent will give us the best option to improve the RF metrics.

$$G_{rf} = I_{dc}^2 \frac{\pi^2 Z_{in} Z_{out}}{V_\pi^2} \tag{2.13a}$$

$$NF_{rf,shot} = \frac{1}{I_{dc}} \frac{2eV_\pi^2}{\pi^2 k_B T Z_{in}} \tag{2.13b}$$

$$SFDR_{imd3,shot} = I_{dc}^{2/3} \left(\frac{2}{e}\right)^{2/3} \tag{2.13c}$$

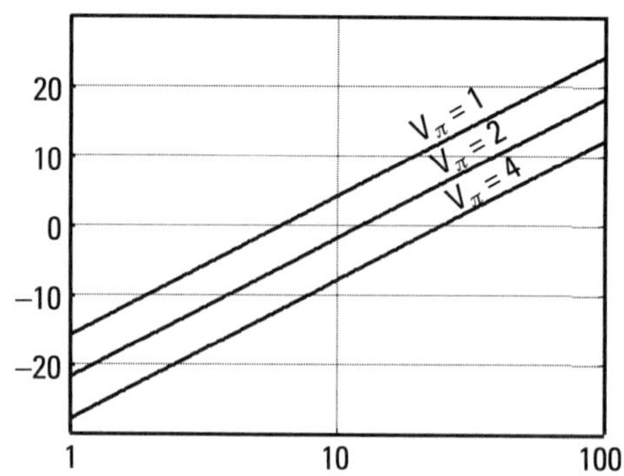

Figure 2.11 RF gain as a function of DC photocurrent for different MZM $V\pi$.

Often there is a question about why the DC photocurrent is the dominant term rather than the RF photocurrent. As seen in Figure 2.10, the majority of the optical power is contained in the optical carrier of the modulated laser and not the sidebands. At the photodetector, the optical carrier mixes with the two sidebands to generate the RF. Thus the optical carrier, which acts like a local oscillator for the mixing at the photodetector, is the dominant term for the RF performance. This leads to the DC photocurrent being the dominant term.

2.4 RF Photonic Component Performance

Having determined the type of modulator to use, the next photonic component to discuss is the laser. The laser provides the optical carrier upon which the RF signal will be upconverted. For an IMDD link, the laser is quantified by two main factors: the output power and the relative intensity noise (RIN). The output power is directly proportional to the DC photocurrent, as the more light that the laser can generate and deliver to the photodetector will both improve the RF gain and NF of the photonic delay line. While the fact that higher RF gain comes with higher optical power intuitively makes sense, the reduction in RF NF is not so obvious. In order to understand this, the relation of shot noise to incoming optical power must be described. Shot noise arises from the random fluctuations in the photocurrent that come from the random arrival time of the photons at the photodetector. The fundamental nature of shot noise is well investigated [14]. For the purpose of this discussion, the important relation to note is that the level of the shot noise in terms of RIN will decrease with increased DC photocurrent [3]. Thus, the RF NF will improve as the DC photocurrent increases. While increasing the DC photocurrent will ideally continuously lower the shot noise level, eventually the RIN of the laser itself will exceed the shot noise limit. The laser RIN will set a lower limit on the level of the noise floor in the system. Table 2.1 shows

Table 2.1
Minimum NFs for Different Laser RIN Levels

Laser RIN (dBc/Hz)	Min RF Noise Figure (dB) at 10 mA
−145	44
−155	34
−165	27
−175	24

the minimum RF NF for different laser RIN levels. Thus, measuring the laser's RIN is very important to predict the RF NF of the photonic link. A plot of the RIN of two different types of lasers appears in Figure 2.12. Some lasers have very low RIN over frequencies as high as 10 GHz. Others have RIN levels that are quite high at lower frequencies. The choice of laser is often determined by the RIN level at a given frequency and the application's needs. The distributed feedback (DFB) semiconductor-based laser in general has the best noise performance over wide frequency bandwidths and is often the preferred solution for most systems. Typical DFB lasers can have RIN levels around –155 dBm/Hz over frequency ranges in the tens of gigahertz. At the same time, they are often limited to optical powers of 100 mW.

In order to increase the optical power while keeping the RIN low, an optical amplifier can be used. Not all RF photonic links require an optical amplifier. A photonic link that requires no optical amplifier is designated as a passive optical network (PON). PONs have been investigated for short- to medium-length links [15]. Examples include fiber-to-the home links as well as some antenna-remoting links. PONs are used when it is impractical to place an amplifier somewhere in the link. Some uses cannot provide electrical power for an optical amplifier, while others cannot afford the monetary cost of an amplifier. In the case of a photonic delay line, amplifiers are required. Even though the loss of the optical fiber is very low, the required time delay for the analog memory application requires very long lengths. For a delay of 100 μs, 20 km of optical fiber is necessary. The minimum optical loss would be 4 dB. From the above equations on RF gain, every 1-dB reduction of DC

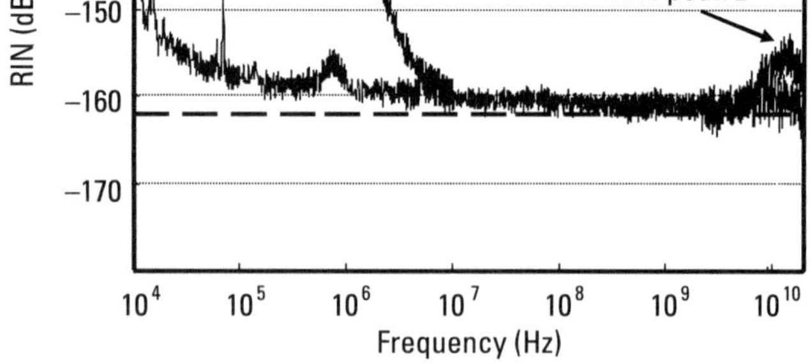

Figure 2.12 RIN of two different types of lasers. (After: [4].)

photocurrent or optical power results in a 2-dB loss of RF power. So in 20 km, the RF power will be decreased by approximately 85%. Therefore the delay line would require optical amplification in order to deliver the appropriate optical power to the end of the delay line. Since all parts of the delay line are accessible, one or more optical amplifiers can be placed at appropriate points in the delay line. The optical gain of the amplifier will then compensate for the loss of the optical fiber. The drawback is that the optical amplifier will add additional optical noise.

While optical amplification can be accomplished by multiple means, the most commonly used amplifier is the EDFA. While many reasons exist to use EDFAs over other types of optical amplifiers [16], the two most important ones are the low optical NF and the extremely slow time response of the gain medium as compared to the other types of amplifiers. The added optical noise of the EDFA affects the RF NF of the overall photonic link. However, the optical NF of the EDFA does not directly translate to the RF NF. A noise penalty metric was defined in order to address how the optical noise adds to the RF NF [17]. EDFAs will typically have a noise penalty between 3 and 24 dB. The result of the noise penalty on the RF NF as a function of DC photocurrent is shown in Figure 2.13. Depending on the application and the noise penalty of the amplifier, the use of an EDFA could provide the necessary optical power for the analog delay line. Unfortunately, the other photonic components will limit the optical power.

After the optical amplifier, the next photonic component is the optical fiber that connects the laser and modulator to the photodetector. As described in Chapter 1, the loss of the optical fiber is already optimized at a value near 0.2 dB/km. After the loss of the optical fiber, the maximum input power the optical fiber can accept is the next important parameter for use in photonic

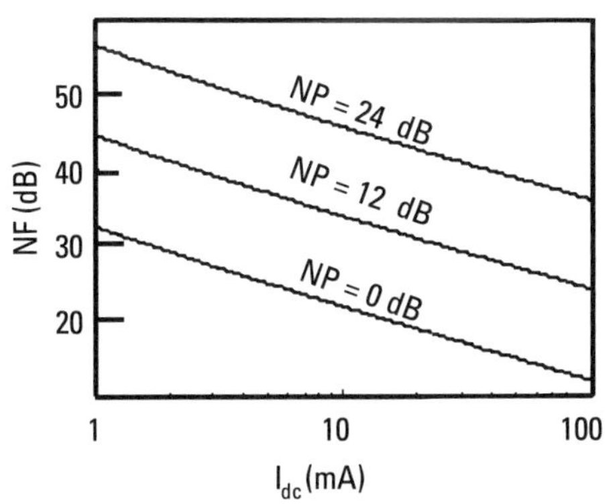

Figure 2.13 RF NF as a function of DC photocurrent at different noise penalties. (After: [4].)

links. As discussed previously, the various RF metrics improve as the optical power delivered to the photodetector increases. However, the optical fiber can only accept a specific amount of optical power before nonlinear optical effects will occur. Once a threshold power is exceeded, the nonlinear optical effects will begin to grow exponentially and affect the performance of the photonic link. The nonlinearity with the lowest threshold power in long lengths of standard optical fibers is the stimulated Brillouin scattering (SBS) effect. The SBS effect is well understood and described in detail in other works [18, 19]. For RF photonic applications, the SBS effect is often characterized by an increase in the reflected optical power at the input of the optical fiber. While Rayleigh scattering will result in some optical power being reflected back to the input of the fiber rather than traveling down to the end of the fiber, it has a linear response as a function of the input optical power. When the SBS threshold is reached, the reflected power increases exponentially as a function of input optical power as seen in Figure 2.14. At the same time, the power received at the end of the fiber no longer increases but flattens as a function of the input optical power, placing a limit on the amount of optical power that can be transmitted down the fiber. The threshold power for the SBS effect is inversely proportional to the length of the optical fiber link. Since analog delay lines require very long lengths of fiber, the SBS threshold can be quite low. While the SBS effect limits the amount of optical power that can be transmitted, it also has another disadvantage. When the SBS threshold is reached, the added noise becomes quite large. The added noise is shown in Figure 2.15. The optical noise is shot noise-limited when the optical power is below the SBS threshold. However, when the SBS threshold is reached, the

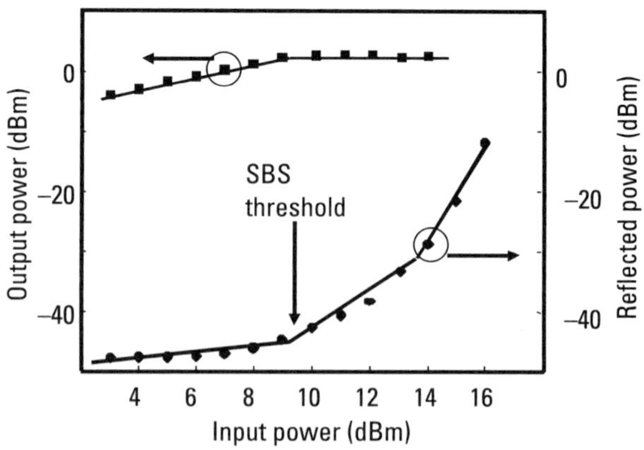

Figure 2.14 Output and reflected power from optical fiber as function of input power. (After: [4].)

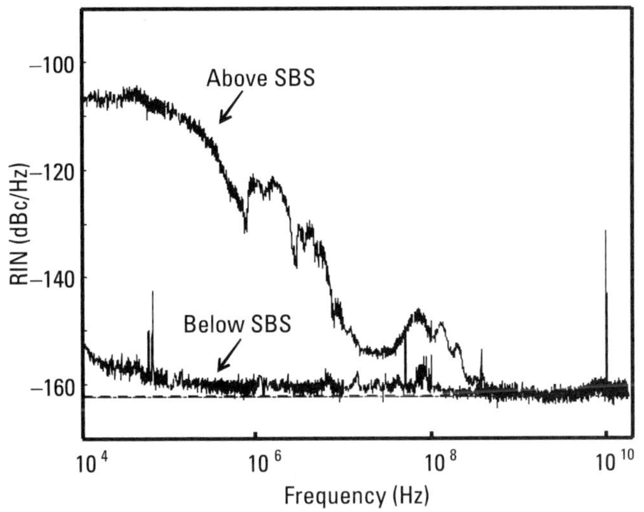

Figure 2.15 Optical noise spectrum above and below the SBS threshold (After: [4].)

added noise level increases significantly. Thus, the SBS effect is a significant problem that has to be designed around.

Another parameter of the fiber link is the chromatic dispersion. The dispersion in optical fibers acts like an optical frequency-dependent time delay, which depends on the length of the fiber. Simply put, different optical frequencies will travel at different speeds through the same length of fiber. In the case of an IMDD link, the optical carrier and the upper and lower sidebands (as seen in Figure 2.10) will all arrive at slightly different times. As the RF frequency increases, the sidebands will spread further apart from each other, causing them to become out of phase with each other. As the phases between the sidebands change, the recovered RF will be degraded. This effect is known as *RF fading* [20]. The results are shown in Figure 2.16, with the RF power shown as a function of RF frequency. For a given length of fiber with a known dispersion constant, the RF power will go from a peak to a null and back again. Much like the phase-modulated link with asymmetric interferometer detection, the chromatic dispersion places a frequency limitation on the link.

An additional effect that occurs due to the chromatic dispersion is the increase in the second-order distortion in the analog photonic link. Normally, the even-order nonlinear distortion is nulled in an IMDD link with an MZM. However, in the presence of chromatic dispersion, the even-order distortions are no longer nulled [21]. The increase in the second harmonic power as a function of frequency is plotted in Figure 2.17. While the fundamental signal

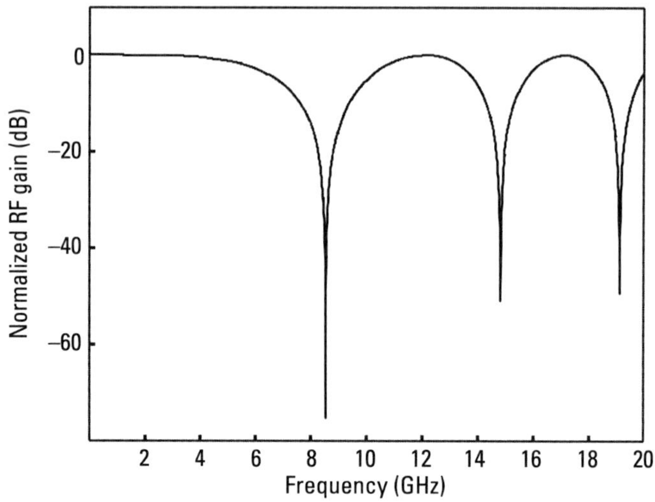

Figure 2.16 RF fading as a function of frequency due to chromatic dispersion. (After: [24].)

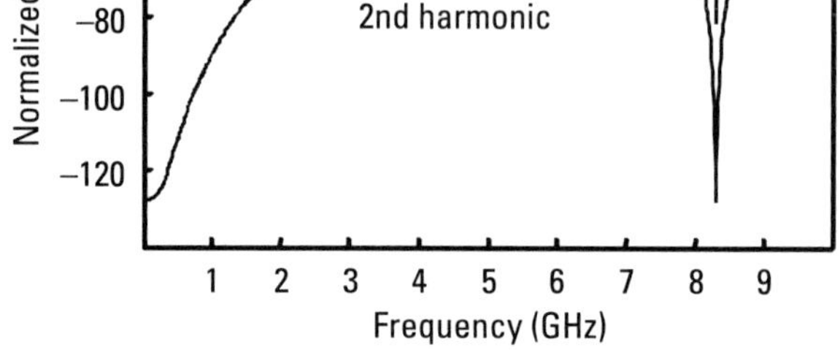

Figure 2.17 Fundamental and second harmonic response due to chromatic dispersion. (After: [4].)

power is not degraded noticeably for lower frequencies, the second harmonic power immediately grows and becomes very significant. Again, this places an operating frequency limitation on the analog photonic link. The chromatic dispersion of the optical fiber link is the cause of both of these deleterious effects. Chapter 3 discusses methods for mitigating this problem.

Finally, the last photonic component is the photodetector itself. The first metric of importance regarding the photodetector is the responsivity. If the slope efficiency of the laser is a measure of the optical power emitted as a function of the drive current, the responsivity of the photodetector is a measure of the photocurrent generated as a function of the optical power input to the device. The higher the responsivity, the more photocurrent that is generated for a given optical power. Just like with the laser, the responsivity is usually less than one. Also just like the laser, the slope is not completely linear. This leads to the photodetector adding spurious signals at the output. This nonlinearity is often overlooked, as the modulator will often add stronger spurs than the photodetector. However, recent work has shown when photonic links can be limited by the photodetector nonlinearities [22].

The other metrics of interest for the photodetector are the optical power handling and the frequency response. Often the power handling and the frequency response cannot be simultaneously optimized. A common reason for this is due to the dimensions of the device. In order to accept higher optical powers, the device often requires a larger absorption region. However, the larger absorption region means that the device has longer transit times for the generated electrons to be swept out of the device. This reduces the frequency response. Newer devices can overcome these limitations, allowing for both higher power handling as well as increased frequency response [23]. These photodetectors can be very useful for many RF applications, as they can meet the required frequency and power handling necessary for delay line applications.

2.5 Conclusions

The analog delay line serves an important purpose in the demonstration architecture. It provides a means for storing the original RF spectrum for future processing needs. Due to the required time delays, an analog delay line is an ideal application for an RF photonic link. While many different versions of a photonic link can be used, the external intensity modulation combined with direct detection is the preferred option. Having determined the type of photonic link that will be used, the RF metrics as a function of the photonic component performance can be defined. This chapter has described the different photonic components metrics in relation to optimizing the performance of the photonic link. Chapter 3 covers advanced techniques for overcoming the individual component limitations. These advanced techniques allow for high-performance RF photonic links to be realized.

References

[1] Friis, H. T., "Noise Figures of Radio Receivers," *Proceedings of the IRE,* Vol. 32, No. 7, 1944, pp. 419–422.

[2] Pozar, D. M., *Microwave Engineering,* Hoboken, NJ: John Wiley and Sons, 1998.

[3] Urick, V. J., et al., "Long-Haul Analog Photonics," *Journal of Lightwave Technology,* Vol. 29, No. 8, 2011, pp. 1182–1205.

[4] Devgan, P., "Review of Radio Frequency Photonics Basics," AFRL Report, AFRL-RY-WP-TR-2017-0157, 2017.

[5] Navid, R., et al., "Close-In Phase Noise in Electrical Oscillators," *Proc. SPIE Symposium Fluctuations and Noise,* 2004, pp. 27–37.

[6] Cox, C. H., et al., "Limits on the Performance of RF-Over-Fiber Links and Their Impact on Device Design," *Transactions on Microwave Theory and Techniques,* Vol. 54, No. 2, 2006, pp. 906–920.

[7] Urick, V. J., et al., "Phase modulation with Interferometric Detection as an Alternative to Intensity Modulation with Direct Detection for Analog-Photonic Links," *Transactions on Microwave Theory and Techniques,* Vol. 55, No. 9, 2007, pp. 1978–1985.

[8] Chi, H., X. Zou, and J. Yao, "Analytical models for Phase-Modulation-Based Microwave Photonic Systems with Phase Modulation to Intensity Modulation Conversion Using a Dispersive Device," *Journal of Lightwave Technology,* Vol. 27, No. 5, 2009, pp. 511–521.

[9] Chain, H.-C., et al., "Optical Millimeter-Wave Generation and Transmission Without Carrier Suppression for Single- and Multiband Wireless Over Fiber Applications," *Journal of Lightwave Technology,* Vol. 28, No.16, 2010, pp. 2230–2237.

[10] Salehi, M. R., and B. Cabon, "Theoretical and Experimental Analysis of Influence of Phase-to-Intensity Noise Conversion in Interferometric Systems," *Journal of Lightwave Technology,* Vol. 22, No. 6, 2004, pp. 1510–1518.

[11] Tkach, R., and A. Chraplyvy, "Phase Noise and Linewidth in an InGaAsP DFB Laser," *Journal of Lightwave Technology,* Vol. 4, No. 11, 1986, pp. 1711–1716.

[12] Cox III, C., *Analog Optical Links,* Cambridge, England: Cambridge University Press, 2004.

[13] Bucholtz, F., et al., "Graphical Approach for Evaluating Performance Limitations in Externally Modulated Analog Photonic Links," *Transactions on Microwave Theory and Techniques,* Vol. 56, No. 1, 2008, pp. 242–247.

[14] Haus, H, A., *Electromagnetic Noise and Quantum Optical Measurements,* New York, NY: Springer, 2012.

[15] Lam, C., *Passive Optical Networks Principles and Practice,* Cambridge, MA: Academic Press, 2007.

[16] Desurvire, E., *Erbium Doped Fiber Amplifiers Principles and Applications,* New York, NY: John Wiley and Sons, 2002.

[17] Urick, V. J., F. Bucholtz, and K. J. Williams, "Noise Penalty of Highly Saturated Erbium-Doped Fiber Amplifiers in Analog Links," *Photonics Technology Letters,* Vol. 18, No. 6, 2006, pp. 749–751.

[18] Kittel, C., *Introduction to Solid State Physics,* New York, NY: John Wiley and Sons, 1996.

[19] Mao, X. P., et al., "Stimulated Brillouin Threshold Dependence on Fiber Type And Uniformity," *Photonics Technology Letters,* Vol. 4, No. 1, 1992, pp. 66–69.

[20] Schmuck, H., "Comparison of Optical Millimeter-Wave System Concepts with Regard to Chromatic Dispersion," *Electronics Letters,* Vol. 31, No. 21, 1995, pp. 1848–1849.

[21] Meslener, G., "Chromatic Dispersion Induced Distortion of Modulated Monochromatic Light Employing Direct Detection," *Journal of Quantum Electronics,* Vol. 20, No. 10, 1984, pp. 1208–1216.

[22] Williams, K. J., "Nonlinear Mechanisms in Microwave Photodetectors Operated with High Intrinsic Region Electric Fields," *Applied Physical Letters,* Vol. 65, No. 10, 1994, pp. 1219–1221.

[23] Williams, K. J., and R. D. Esman, "Design Considerations for High-Current Photodetectors," *Journal of Lightwave Technology,* Vol. 17, No. 8, 1999, pp. 1443–1454.

3

Advancements in Analog Delay Line Performance

Chapter 2 identifies the analog buffer as one of the applications where RF photonics can provide an advantage over other solutions. Specifically, an RF photonic link can be used as an analog delay line, allowing for the storage of a wide spectrum of signals for up to hundreds of microseconds. In fact, Chapter 2 makes the case for using an IMDD link as the preferred solution for the delay line and defines the RF metrics for this architecture as a function of the photonic components' performance.

Given the definition of IMDD link performance, the only way to improve the performance is to increase the optical power, decrease the electro-optic conversion efficiency (measured by the V_π), or reduce the optical noise from the laser (and optical amplifier if used). However, these actions all require improving the individual photonic component performance. Perhaps there are other means to improve the performance without changing the individual photonic components? This chapter presents advanced techniques to improve the RF performance of the analog delay line without fundamentally changing the individual components. In doing so, this chapter provides RF photonic link designers with additional tools to use when building a potential system. While the focus is on the analog delay line, the techniques presented here will also improve any RF photonic link.

3.1 Performance Improvement Through the Photonic Components

The RF metrics are given in the logarithmic scale in (3.1a)–(3.1d), where V_π is the half-wave voltage of the MZM in units of volts, $I_{\rm dc}$ is the DC photocurrent in units of milliamps, and $Z_{\rm in}$ and $Z_{\rm out}$ are the input and output impedance, respectively. The assumptions for these expressions are that the MZM is biased at the quadrature bias point, that the photodiode is completely linear in its response, and that the input and output impedances of the MZM and the photodetector are 50Ω. If the shot noise is the dominant source, the RF metrics can be further simplified. Equation (3.2a) is the shot noise contribution in logarithmic scale. Substituting (3.2a) into (3.1b)–(3.1d) yields the simplified forms shown in (3.2b) and (3.2c), with e as the electron charge. From (3.1a), the RF gain is dependent on two parameters, the DC photocurrent ($I_{\rm dc}$) from the photodetector and the V_π of the MZM. Similarly, the RF noise figure is dependent on at least one of the two parameters. With the above assumptions, one simple way to improve the RF metrics is simply to increase the optical power delivered to the photodetector. Examples of RF photonic links with improved RF gain and NF have been previously demonstrated [1–3]. The increased optical power can be accomplished by using multiple lasers as well as using high-power optical amplification to increase the total power delivered to the photodetector. While the increased optical power will improve the RF metrics, limitations exist on how much improvement can be realized.

$$G_{\rm rf}\,[\rm dB] = -22 + 20\log\left(I_{\rm dc}\,[\rm mA]\right) - 20\log\left(V_\pi\,[\rm V]\right) \tag{3.1a}$$

$$NF_{\rm rf}\,[\rm dB] = 174 - G_{rf} + N_{\rm total}\,[\rm dBm/Hz] \tag{3.1b}$$

$$OIP_3\,[\rm dBm] = -13 + 20\log\left(I_{dc}\,[\rm mA]\right) \tag{3.1c}$$

$$SFDR_3\,[\rm dB] = \tfrac{2}{3}\left(OIP_3 - N_{out}\right) \tag{3.1d}$$

$$\begin{aligned} N_{\rm shot}\,[\rm dBm/Hz] &= 10\log_{10}\left(0.5eI_{dc}Z_{out}\right) \\ &= -174 + 10\log_{10}(I_{dc}\,[mA]) \end{aligned} \tag{3.2a}$$

$$NF_{\text{rf}}\,[\text{dB}] = 22 - 10\log\left(I_{\text{dc}}\,[\text{mA}]\right) + 20\log\left(V_{\pi}\,[\text{V}]\right) \tag{3.2b}$$

$$SFDR_3\,[\text{dB}] = \tfrac{2}{3}\left[161 + 10\log\left(I_{\text{dc}}\,[\text{mA}]\right)\right] \tag{3.2c}$$

3.2 Improvements in the Photodetector

One of the first limitations is on the photodetector itself. Photodetectors with a frequency response above 20 GHz can operate linearly when the input optical power is kept at 5–10 mW or lower. Beyond this optical power, the typical photodetector will no longer operate with a linear response and will often be destroyed. A common failure mechanisms is overheating due to the intense amount of light being focused on the small active region of the device. To overcome this problem, new designs for photodetectors have been developed. One method involves the use of a gradient index (GRIN) lens before the photodiode to distribute the optical power more uniformly over the active region [4]. This allows the photodetector to handle more optical power before the heating problem destroys the device. Other photodetectors with a frequency response above 20 GHz and operating input powers in the 100-mW range have been shown [5–8]. Using a high-power handling photodetector, an RF link with RF NFs below 10 dB at frequencies as high as 10 GHz being achieved [9].

While photodetectors are often modeled as linear devices, they do have nonlinear responses that can adversely affect the RF metrics of an analog delay line. One way that photodetectors can adversely affect analog delay lines is in the phase noise performance. Recall from Chapter 2 that some analog delay lines can be used to model long-distance transmission and that the added phase noise needs to be minimized. Previous work in distributing and generating low-phase noise clock signals using photonics have shown that the photodetector can be the limiting component, due to intensity-to-phase noise conversion in the photodetector [10–12]. In order to minimize this effect, a method has been proposed to optically saturate the photodetector [13]. By compressing the photodiode, the intensity-to-phase response can be clamped and thus the added phase noise can be reduced. The results of the phase noise performance when the photodiode is saturated or not appears in Figure 3.1. By correctly choosing the saturation point of the photodiode, the phase noise is improved by over 10 dB in the 10 kHz–1 MHz frequency

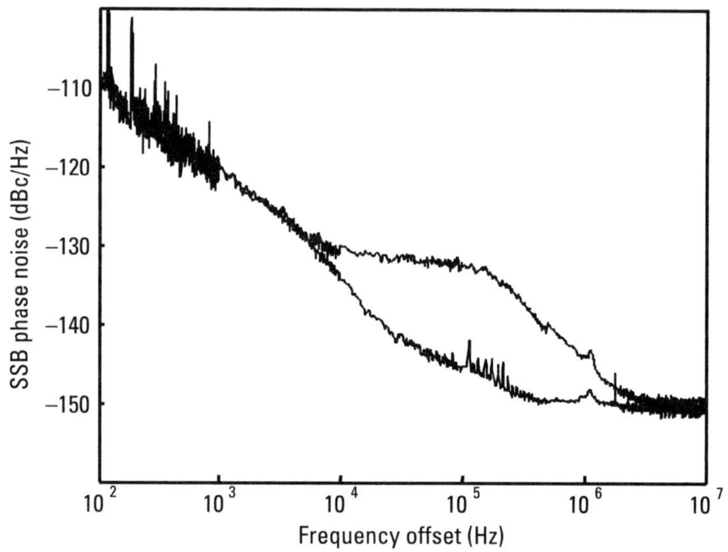

Figure 3.1 Phase noise of a photodetector in and out of saturation. (After: [14].)

offset range. When used in a 14-km long delay line, the overall timing jitter is within 0.3 fs (femtoseconds) of the case where there is no delay line. The trade-off for saturating the photodetector is a reduction in available RF power from the output of the device. A method for overcoming this drawback is addressed in Section 3.5.2.

Along with the added phase noise, the response of the photodetector does add second- and higher-order nonlinear distortions. The nonlinearity originates from the voltage dependent responsivity of the photodetector and has been studied [15]. Work has been done to make photodetectors with better linearity, with reports of measured output intercept points of the second harmonic ($OIP2_{2H}$) of 50 dBm [15, 16]. However these are noncommercial, specialty photodetectors, and the measurements were limited to the 1–2-GHz frequency range. These effects can limit the overall RF performance of the analog delay line and have to be considered in designing the delay line. The required nonlinear performance for the photodetector has been investigated previously [17]. In order to overcome the third-order nonlinearities due to the photodetector, predistortion techniques have been used at the modulator [18–20]. Often, these involve added electronic components and come at the cost of a reduction in the power of the fundamental RF frequency. A photonic configuration for canceling the second-order distortion of the photodiode is discussed in detail later in Section 3.5.5.

While work continues to address the deficiencies of single photodetectors, the analog delay line performance can be improved through the use of balanced detection. Balanced photodetection can provide several benefits. If one photodetector can handle a certain amount of optical power, then two photodetectors can handle twice as much optical power. For example, the increase in RF gain can be 6 dB, as has been shown in phase-modulated links combined with balanced detection [21]. While balanced photodetection is common for phase-modulated links, it can be used in intensity-modulated links as well. Since the MZM intensity modulator can be made with two optical outputs, each output can be then connected down an independent fiber to the inputs of the balanced photodetector. The resulting link would provide the same increased RF gain as the phase-modulated link. Demonstrations of links using parallel optical fiber paths have been shown in the lab [22]. Note that the two fiber-optic paths have to be length-matched to less than a single wavelength of the highest frequency at which the link will operate. While such a link could be hard to realize in real-world environments, there have been demonstrations using multiple parallel links combined with balanced photodetection [23]. Further studies have also shown that colocated optical fibers can have relatively small phase changes over large temperature ranges, especially if they are buried [24]. For the analog delay line, the temperature and fiber lengths can be properly controlled to make use of the benefits of balanced detection.

Balanced photodetection can also provide a way to reduce the impact of laser RIN on the RF metrics. Note that the two optical outputs of the MZM are 180° out of phase with each other. When the two outputs are connected to the balanced photodetector, the photocurrent generated from the individual photodetectors is also 180° out of phase with each other. The result is that the recovered RF signal now adds coherently. On the other hand, the laser noise is in-phase on the two MZM outputs. Therefore, the noise will not add in the balanced photodetector but will instead be subtracted. This is known as common-source RIN rejection [25]. So not only does the RF gain increase, the noise is simultaneously suppressed. The amount of noise that is suppressed depends on the source and how well matched the two paths are. For analog delay lines, this type of architecture can help bypass the limits on RF NF that the RIN of the laser imposes. Since the RIN is from a common source in both paths, it can be suppressed. Balanced photodetection has been well investigated for improving RF NF [26].

In addition, the use of balanced photodiodes can cancel the even-order distortion due to the individual photodiodes [22]. However, each photodiode will react differently and the cancelation will not be optimized when the bias voltage of each photodiode is the same. In practice, each photodiode's bias is set to optimally cancel the even-order distortion. The optimal bias voltage depends on the analog delay line architecture. Using the dither tone from the MZM bias controller, the bias voltages of each of the photodiodes can be automatically tuned to minimize the even-order distortion due to the photodiodes. A demonstration using a microcontroller to optimize the bias voltages showed an improvement of 8.3 dB in the second harmonic distortion, while the fundamental RF power, third harmonic distortion, and RF NF are not degraded [27].

The system works by using a microcontroller to scan the optimal bias voltage in order to minimize the second harmonic of the dither tone. Figure 3.2 shows the power of the second harmonic as the bias voltage changes. After the minimum second harmonic power is found, the microcontroller sets the correct bias voltages for the photodiodes. Using the equation for the required OIP2 in order to remain third-order limited $OIP2_{req}(dBm) = 47 + 70/3 * \log_{10}(I_{dc}(mA))$ [27], the achieved OIP2 can provide third-order limited performance for photocurrents up to $I_{dc} = 33.5$ mA.

3.3 Improvements in the Optical Fiber

With multiple methods to overcome the limitations of the photodetector, the next component that needs to be overcome is the optical fiber. Due to the nonlinear SBS effect, incoming light above the SBS threshold will limit the light that can be sent down the fiber, reflecting any extra light back to the input. The dispersion in the optical fiber will also add impairments to the RF

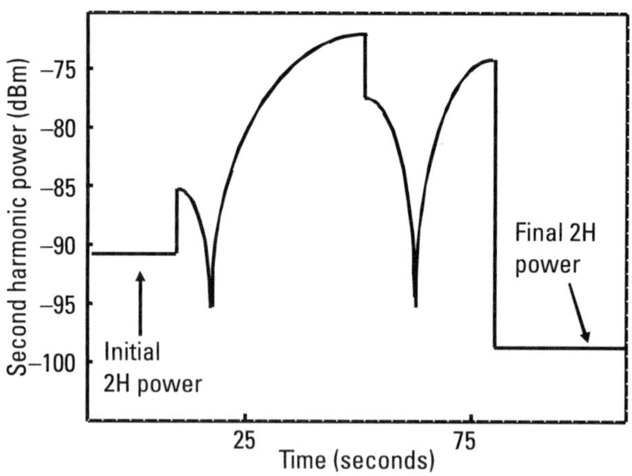

Figure 3.2 Improvement in second harmonic changing the bias voltage. (After: [14].)

signal. To combat both of these problems, multiple solutions exist. A dispersion compensation fiber module can be used at the end of the optical fiber to correct the impairments to the original signal. Such modules are common in long-haul digital fiber-optic networks [28–31]. However, they add loss and do not address the SBS effect. To increase the SBS threshold in optical fibers, different approaches have been introduced, ranging from changing the temperature of the fiber [32], introducing strain [33], and adding another waveguide to decouple the acoustic traveling wave from the optical light [34]. Changing the temperature of the fiber can be done in a laboratory environment, but it is not very practical for use in real-world applications. Fabricating custom fibers increases the cost and may only be produced in limited lengths.

One method that has been explored to increase the amount of power before triggering the SBS effect is the use of alternating pairs of fiber spans with different SBS spectra [35, 36]. Unfortunately, the increase is only around 2 dB as compared to a single fiber type. To increase the input power even further, optical isolators can be placed between each alternating pair. The isolators act to make each alternating pair of fibers independent from each other. Moreover, by choosing two optical fibers with opposite dispersion signs, the overall dispersion of the fiber link can be made close to zero. As shown in Figure 3.3, an example link shows the use of an optical fiber with a dispersion of –4ps/(nm•km) and another optical fiber with a dispersion of +8ps/(nm•km). By using a length of fiber with the lower dispersion that is twice the length of the higher dispersion fiber, the overall fiber will have a dispersion close to zero. The combination of fibers with the isolators also increases the SBS threshold. As seen in Figure 3.4(a), the alternating fiber with optical isolators increases the SBS threshold by 6 dB when compared to a single optical fiber span. In tandem, the optical power that can be transmitted down the fiber is increased by 6 dB when compared to a single fiber, as illustrated in Figure 3.4(b). Thus, the alternating fiber span addresses both the dispersion and the SBS issues. The improvements do come at a cost. Overall fiber losses increase by adding the isolators and splicing the different types of fiber together. In addition, the configuration cannot be used when access to the fiber is not available.

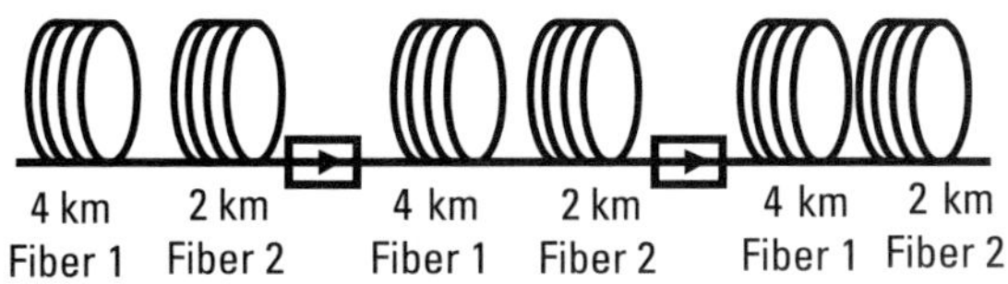

Figure 3.3 Optical fiber span with isolators in-between. (After: [14].)

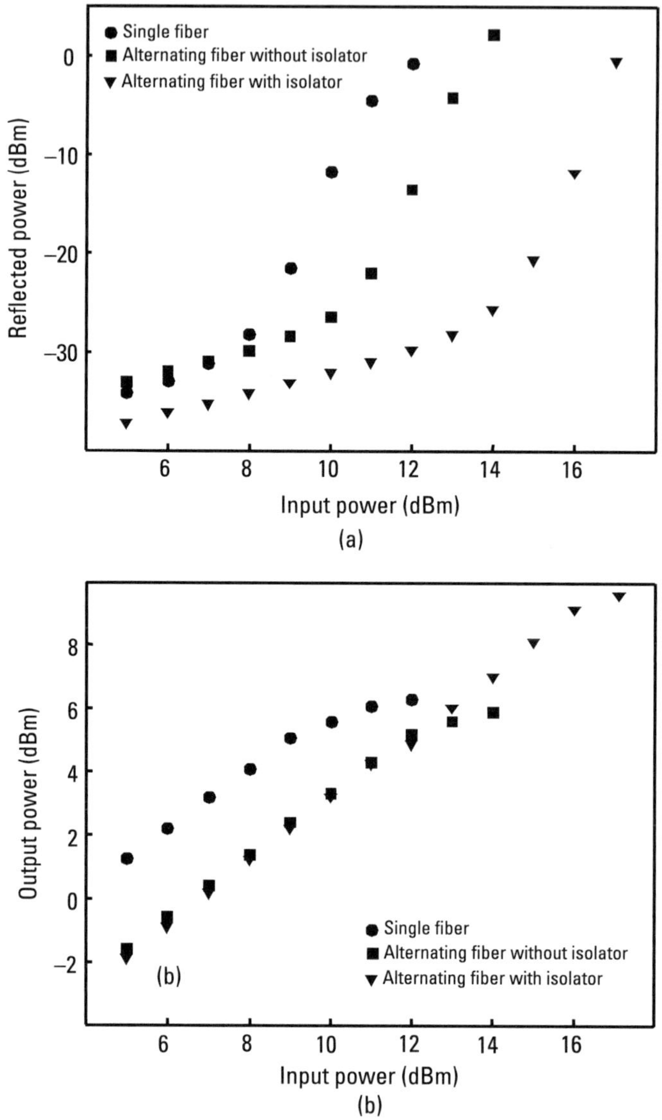

Figure 3.4 (a) Measure of the reflected optical power below and above the SBS threshold and (b) transmitted optical power below and above the SBS threshold. (After: [14].)

3.4 Improvements in the Optical Amplifier

The optical amplifier is the next component that needs to be addressed. The optical amplifier can take the optical power from a low-power laser and increase it significantly. The increase in power comes at a cost. One detriment that is often overlooked is the addition of spurious signals. Optical amplification systems that have very fast response times can have their gain modulated by the RF signal that is present on the incoming optical light that is to be amplified. While these effects can be useful for other applications [37, 38], they seriously degrade the performance of analog systems. Semiconductor- or nonlinear-based optical amplification (e.g., four wave mixing) has response times ranging from picoseconds to femtoseconds. The EDFA, on the other hand, has response times on the order of milliseconds, making it ideal for analog photonic links [39]. The EDFA is the preferred optical amplifier for RF photonic applications.

Another cost to using an optical amplifier is the added noise. The added noise from the EDFA will degrade the RF metrics of the photonic link. While the EDFA's optical NF is well understood [40], the effect the optical NF has on the RF metrics is not as clearly defined. In order to address this issue, the RF noise penalty metric was developed [41]. The noise penalty of the EDFA is measured using a similar method to measuring the RIN of a laser. The power spectral density is recorded and compared to the shot noise limit for the operating photocurrent. The difference is then called the noise penalty. An example measurement is shown in Figure 3.5. The noise penalty will affect the RF metrics as shown in (3.3a)–(3.3c), where $\varphi_{1\text{dB}}$ is the RF phase shift in the MZM which drives the link's output power 1 dB into compression and RIN_{sh} is the shot noise relative to the photocurrent. The RIN_{sh} is the dominant noise term for photocurrents between 1 and 100 mA. For a delay line of 100 μs or more, multiple EDFAs could be used in the link. To calculate the overall noise penalty from multiple EDFAs, a cascaded noise penalty formula was derived [42]. The cascaded NF depends on the gain of each EDFA, the

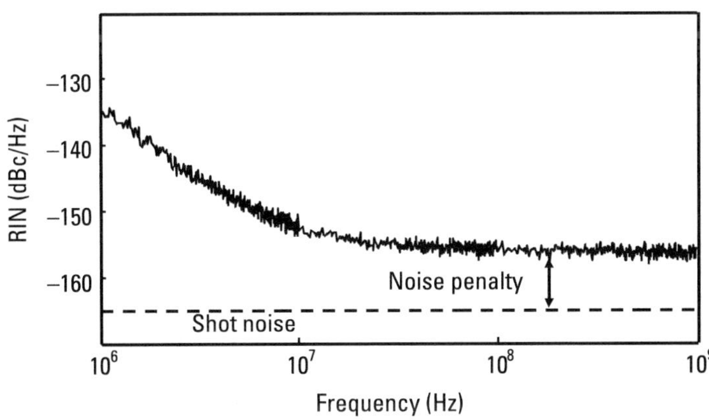

Figure 3.5 Noise penalty of an EDFA. (After: [14].)

noise penalty of each EDFA, and the loss between each EDFA. An example link is shown in Figure 3.6, along with the cascaded noise penalty equations in (3.3a)–(3.3c). Equation (3.4) has the overall noise penalty as a function of l_o, the loss before the photodetector, L_n; the optical loss of each length of fiber, G_n; the gain of each EDFA; and NP_n, the noise penalty of each EDFA. The measured noise penalty of the link was shown to be in very good agreement with the predicted noise penalty from the cascaded noise penalty equation.

$$NF_{rf,\,amp} = NP \cdot NF_{rf,\,sh} = \left(\frac{(RIN_{sh})V_{\pi}^{2}}{\pi^{2} k_{B} T Z_{\text{in}}} \right) NP \tag{3.3a}$$

$$CDR_{amp} = \frac{CDR_{sh}}{NP} = \frac{\varphi_{1\text{dB}}^{2}}{2(RIN_{sh})(NP)} \tag{3.3b}$$

$$SFDR_{amp} = \frac{SFDR_{sh}}{NP^{2/3}} = \left(\frac{4}{(RIN_{sh})(NP)} \right)^{2/3} \tag{3.3c}$$

$$RIN_{sh} = \frac{2q}{I_{dc}} \tag{3.3d}$$

$$\begin{aligned} NP_c \approx 1 + l_o[L_2 \bullet\bullet\bullet G_n(NP_{EDFA,\ 1} - 1) + \ldots\ldots \\ + L_n G_n (NP_{EDFA,\ n-1} - 1) + (NP_{EDFA,\ n} - 1)]\ . \end{aligned} \tag{3.4}$$

With a method for determining the amount of noise penalty the EDFA will introduce to the RF photonic link, the next step is to determine the best operating point for the EDFA. Previous work has shown that an EDFA

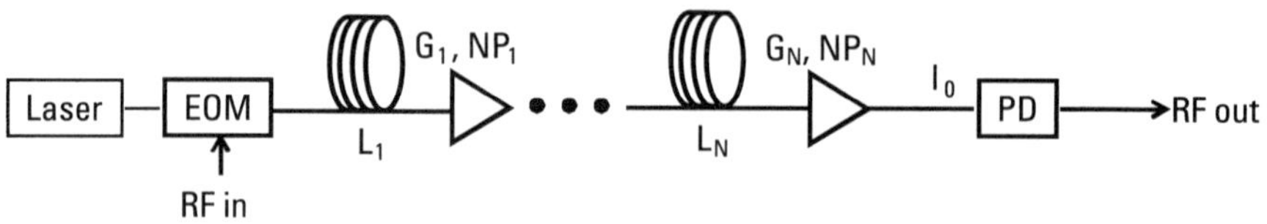

Figure 3.6 Noise penalty link. (After: [14].)

operating in saturation will reduce the noise penalty significantly, due to the reduction in spontaneous emission [43]. The EDFA will enter saturation as the optical power at its input increases. The noise penalty of the EDFA will decrease as the amplifier goes further into compression. An example of the noise penalty of an EDFA as a function of input power is shown in Figure 3.7. Note that the trade-off is that an EDFA in compression will not provide as much optical gain as when it is operating in the linear regime.

3.5 Improvements in the Optical Modulator

Having addressed the EDFA, optical fiber, and photodetector, the next step is to look at the optical modulator. The key metrics are the insertion loss and the modulator efficiency (V_π). The insertion loss will limit the amount of optical power that appears at the output of the modulator. This in turn will reduce the amount of photocurrent at the output of the analog delay line. In addition, the reduced output power will make it harder to saturate the EDFA if it is placed after the MZM. So the insertion loss is just as important as the V_π of the modulator. The V_π is dependent on both the material that is used along with how well the RF electric field is matched to the optical field [44]. Since both the V_π and insertion loss of the device are set at the time of manufacture, it cannot be changed by the analog delay line designer. Instead, the MZM can be used in different ways to overcome the V_π and insertion loss.

3.5.1 Off-Quadrature Biasing of the Optical Modulator

Another method for operating the MZM to improve the RF metrics is to change the DC operating bias point. As stated previously, the quadrature bias

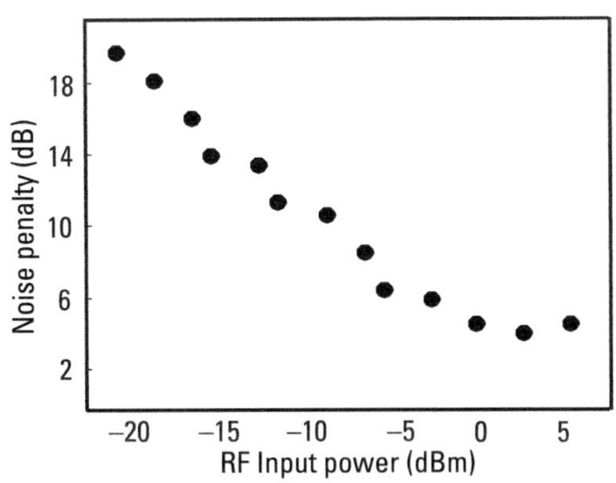

Figure 3.7 Noise penalty as a function of the input power to the EDFA. (After: [14].)

point is the preferred operating point for an MZM intensity-modulated link. For an unamplified link, quadrature-biasing results in the highest RF gain as well as providing the conditions for canceling the even-order distortions after photodetection. However, in a link with an MZM followed by optical amplification, the quadrature bias may no longer provide the highest RF gain. In fact, the highest RF gain occurs at a point below the quadrature bias point. This is often referred to as low-biasing of the MZM. The exact operating point is different for each MZM, as it depends on the extinction ratio and V_π of the MZM along with the gain and P_o, the output saturation power of the EDFA. The relation between RF gain and I_{dc} and V_π appears in (3.5), where R is the responsivity of the photodetector.

$$G_{rf} = I_{dc}^2 \frac{\pi^2 \sin^2(\varphi_{dc}) Z_{in} Z_{out}}{V_\pi^2 \left[1 + \left(\frac{I_{dc}}{RP_{o.\max}} \left[1 - \cos(\varphi_{dc}) \right] \right)^\gamma \right]^2} \tag{3.5}$$

The increased RF gain for operating at the low bias point comes from the improved optical gain of the amplifier out of compression. As seen in Figure 3.8, when the MZM is operated at quadrature bias, the unmodulated optical carrier is much stronger than the sidebands that carry the RF signal. After passing through the EDFA, the strong optical carrier compresses the EDFA and steals the majority of the optical gain, keeping the modulation depth low. On the other hand, when the MZM is operated at the low bias point, the optical carrier is suppressed to be about the same power as the sidebands. The EDFA is no longer in compression, and the available optical

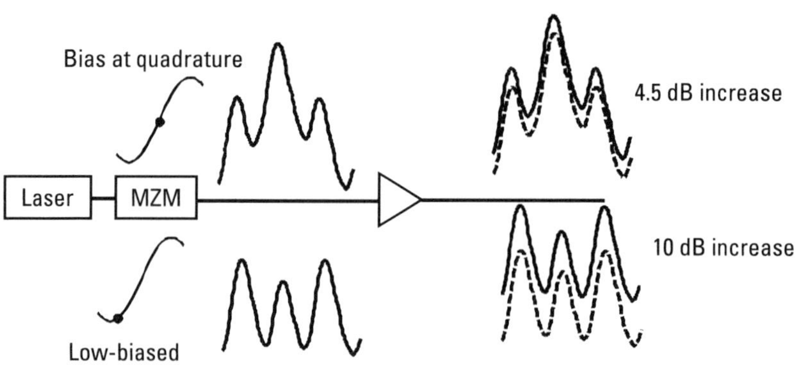

Figure 3.8 Optical carrier and sidebands before and after EDFA at quadrature and low bias.

gain is higher. After passing through the EDFA, the carrier (along with the sidebands) experiences an increase in the modulation depth. This in turn increases the RF gain of the link. The gain improvement is referred to as all-photonic gain, as it does not depend on any electronic amplification to achieve higher RF power.

Multiple advantages come with operating the MZM at the low bias point. Previous work has shown that both the RF gain and the RF NF can be improved when compared to a quadrature bias MZM at the same photocurrent [45]. The improvement in the RF gain is shown in Figure 3.9. As the bias voltage goes below quadrature ($V_{\text{bias}} = 0.5^* V_\pi$), the modulation index, along with the RF gain, increases. It peaks and then quickly drops as the carrier becomes completely suppressed. The increase in RF gain is 10 dB as compared to quadrature bias. Note that since the optical carrier is suppressed, the DC photocurrent is much lower than at the quadrature bias point. While the photocurrent is lower, the higher modulation index may still lead to compression of the photodetector. Additionally, by suppressing the optical carrier, the delay line can operate below the SBS threshold of the optical fiber while still providing the necessary RF performance.

3.5.2 Low Biasing of the Optical Modulator with Dual Wavelengths

As mentioned previously, operating a photodiode in saturation is beneficial for certain analog delay lines. Compressing the photodiode clamps the intensity-to-phase response and reduces the added phase noise. Unfortunately, running the photodiode in compression limits the RF power available at the

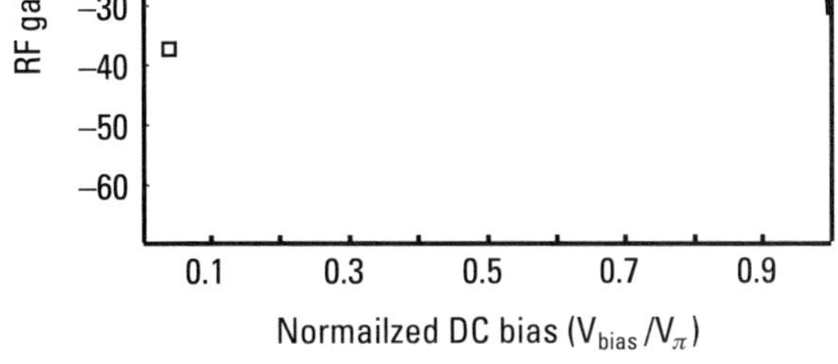

Figure 3.9 RF gain as a function of DC bias voltage relative to V_π. (After: [14].)

output of the analog delay line. To overcome this limitation, an RF amplifier can be used to reach the desired power. Unfortunately, the flicker noise of the RF amplifier will place a lower bound on the phase noise of the signal [46]. By using a low-biased MZM followed by and EDFA, the all-photonic gain can provide the necessary RF output power, even with the photodetector in saturation. Combining the all-photonic gain with a saturated photodetector provides an analog delay line with low phase noise and higher RF output power.

Using a low-biased MZM followed by an EDFA does come at a cost. Since the MZM is no longer operating at the quadrature bias point, the even-order distortion is no longer canceled at the photodiode. The resulting even-order distortion can actually be quite large and limits these types of delay lines to single-octave operation. Another disadvantage comes from the fact that the EDFA is no longer operating in the saturation regime. While this provides more gain to the sidebands, it also increases the RF noise penalty associated with the EDFA, as mentioned previously. The added noise penalty can be detrimental to the overall RF NF of the delay line. The EDFA has to be carefully selected in order to ensure that the added noise does not overcome the improvement from low biasing.

In order to overcome some of these limitations, an advanced form of low biasing was developed [47]. Other demonstrations have been made with multiple wavelengths in order to improve the RF metrics of photonic links [48]. These other demonstrations used only a single-output MZM. An advanced system is shown in Figure 3.10. The technique uses a dual-output MZM with two wavelength inputs, one operating at low bias and the other operating at high bias. Since the MZM has a wavelength-dependent electro-optic coefficient, the transfer curve of the modulator as a function of DC bias voltage is different for various wavelength inputs. Thus, for a given DC bias voltage applied to the MZM, different wavelength inputs will operate at different DC phase shifts. The result is illustrated in Figure 3.11. By using a dual-output MZM, each wavelength can then be low-biased at their respective output. The first transfer curve is for one wavelength while the other

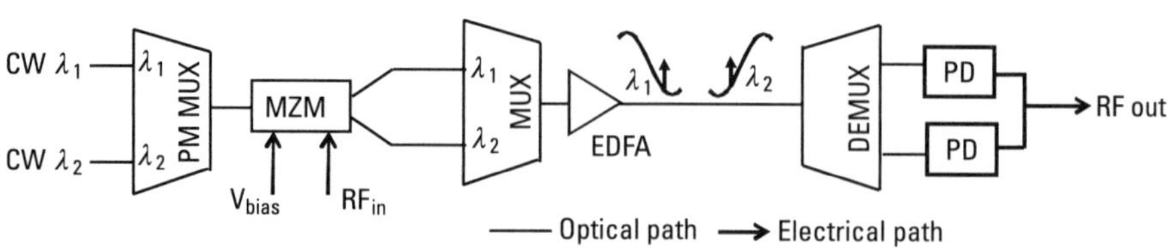

Figure 3.10 RF subsystem using two low-biased wavelengths to remove even order distortion. (After: [14]).

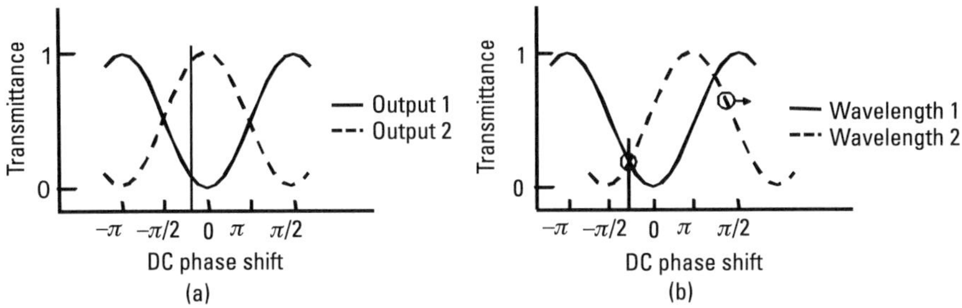

Figure 3.11 Operating two wavelengths each at low bias. (After: [14]).

transfer curve is for a second independent wavelength. By wavelength multiplexing the low-biased first wavelength from one arm and the complementary low-biased second wavelength from the other arm, one can transmit two low-biased wavelengths to the optical amplifier. The amplifier output is then sent to a wavelength demultiplexer and the separated wavelengths are each sent to a photodiode input. The outputs of each photodiode are then subtracted in order to recover the RF input. Combining the low-bias outputs through an optical amplifier can simultaneously improve the various RF metrics as compared to a quadrature-biased MZM through the same optical amplifier.

$$I_{output1}\left(\varphi_{\mathrm{dc},\lambda_1},t\right) \propto 1-\cos\left(\varphi_{\mathrm{dc},\lambda_1}\right) + \sin\left(\varphi_{\mathrm{dc},\lambda_1}\right)\sin\left(\Omega_{rf}t\right) - \cos\left(\varphi_{\mathrm{dc},\lambda_1}\right)\cos\left(2\Omega_{rf}t\right) \tag{3.6a}$$

$$I_{output2}\left(\varphi_{\mathrm{dc},\lambda_2},t\right) \propto 1+\cos\left(\varphi_{\mathrm{dc},\lambda_2}\right) - \sin\left(\varphi_{\mathrm{dc},\lambda_2}\right)\sin\left(\Omega_{rf}t\right) + \cos\left(\varphi_{\mathrm{dc},\lambda_2}\right)\cos\left(2\Omega_{rf}t\right) \tag{3.6b}$$

$$\varphi_{dc,\lambda_2} = -\pi - \varphi_{dc,\lambda_1} \tag{3.6c}$$

$$RIN_{sh} = \frac{2q}{I_{dc}} \tag{3.6d}$$

Equations (3.6a) and (3.6b) show the photocurrent for each wavelength at the output of the demultiplexer, with φ_{dc} as the DC bias point for each wavelength. By setting the correct phase bias condition between the two wavelengths (3.6c), the even order distortion will be cancelled while the improved fundamental is preserved (3.6d). The advantage of the dual output MZM method is it requires a smaller wavelength separation for phase bias φ_{dc} greater than 0.25π output MZM method. For a specific instantiation of this technique, the wavelength separation was only 5 nm as compared to 10 nm in the single output case. With two wavelengths, twice the optical power is input to the EDFA. This pushes the EDFA further into compression, reducing the noise penalty. The dual wavelength, low biased setup improves the RF gain, noise figure and spur free dynamic range when compared to a quadrature biased link operating at the same DC photocurrent. In addition, since the optical carriers are suppressed, the link length can be increased by a factor of five before the SBS threshold is reached. The overall improvements in the performance of the RF metrics are shown in Table 3.1 as compared to an MZM link biased at quadrature and operating at the same photocurrent. Thus the use of multiple wavelengths can improve all the RF metrics.

This technique does have some drawbacks. It requires two lasers, which adds to the complexity of the system. It also requires path matching of the two outputs from the MZM before they are combined at the EDFA. Finally, while the system uses balanced detection, the two lasers are not common so the noise will not be cancelled. Also, even though the EDFA is common, the filtered noise is not from the same part of the spectrum and thus is also not common. Nevertheless, this technique points to the use of multiple wavelength sources as a means to overcome some of the penalties of using non-quadrature biased modulation.

3.5.3 Cancelation of Dispersion Induced Second Harmonics by Using Dual Wavelengths

The dual laser system with a single output has also been shown to compensate for the second harmonic distortion that is introduced by dispersion in optical fibers. Chromatic dispersion has been shown to limit the frequency bandwidth of fiber links, with a typical 50 km link having a 3-dB bandwidth of ~7 GHz [49]. While this limit is significant, the dispersion-induced second harmonic is even more restricting. The dispersion-induced second harmonic can limit the analog performance bandwidth as low as 3 GHz for a 50-km link [50]. Clearly, this limitation is significant for long-haul microwave frequency analog links. While the alternating fiber pair configuration can be used to remove the dispersion penalty, the availability of the right fiber types may not be possible. This technique can be used for these situations.

The technique of using a dual-input MZM with two separate wavelength inputs, which cancels the chromatic dispersion-induced second

Table 3.1
RF Metric Improvements from the Two Laser Low Bias System versus Quadrature Bias at the Same Photocurrent

	Improvments in Link Metrics
RF Gain	9 dB
RF NF	5.4 dB
RF SFDR	3.6 dB
Length	5x increase

harmonic, appears in Figure 3.12. Note that the architecture is almost the same as the two laser system in Table 3.1, with two exceptions. First, there is no EDFA, and second, the lasers are not set for low bias but for quadrature bias. In this application, the unamplified quadrature-biased performance is sufficient. Again using the MZM's wavelength-dependent electro-optic coefficient, two wavelengths are chosen to operate at the same quadrature point of their respective transfer curves. The output of the MZM is then passed through a length of fiber with high chromatic dispersion. After the fiber, a WDM separates the two wavelengths, and each is sent to a photodiode input. The outputs of each photodiode are then subtracted in order to recover the RF input. The results show a cancelation of the second harmonic while also increasing the RF power of the fundamental signal by 6 dB as compared to a single wavelength.

Equations (3.7a) and (3.7b) show the proportional received photocurrent due to an optical carrier modulated by an RF signal via a MZM biased at quadrature and after passing through a fiber with nonzero chromatic dispersion. The equations have been truncated to the DC (first term), fundamental (second term), and the second harmonic (third term) terms for an optical carrier that we have designated as λ_1 and λ_2, respectively. Using the

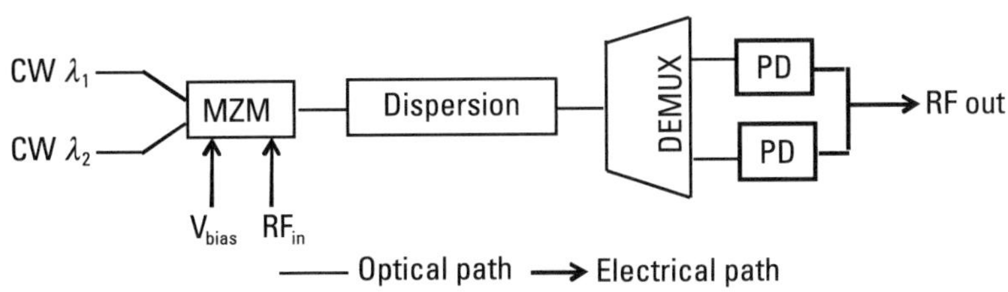

Figure 3.12 System for canceling the second-order distortion due to chromatic dispersion. (After: [14].)

small-signal approximation of $OIP2_{req}(dBm) = 47 + 70/3 * \log_{10}(I_{dc}(mA))$, with φ_{rf} as the phase shift in the MZM arms due to the RF signal yields

$$I_{\lambda 1}(\Omega,t) = I_{DC} + I_{DC}\varphi_{rf}\cos(\frac{1}{2}\theta_2)\sin(\Omega t - \theta_1) + \frac{I_{DC}\varphi_{rf}^2}{4}\cos(2\theta_2)\cos(2\Omega t - 2\theta_1) \tag{3.7a}$$

$$I_{\lambda 2}(\Omega,t) = I_{DC} - I_{DC}\varphi_{rf}\cos(\frac{1}{2}\theta_2 + \varphi_{\Delta\lambda})\sin(\Omega t - \theta_1 + \varphi_{\Delta\lambda}) + \frac{I_{DC}\varphi_{rf}^2}{4}\cos(2\theta_2 + 2\varphi_{\Delta\lambda})\cos(2\Omega t - 2\theta_1 + 2\varphi_{\Delta\lambda}) \tag{3.7b}$$

where $\theta_1 = \beta_1\Omega L$, $\theta_2 = \beta_2\Omega^2 L$, β_1 is the phase velocity of each optical carrier, β_2 is the group velocity dispersion of each optical carrier, Ω is the angular RF frequency, L is the length of the fiber between the output of the MZM and the photodiode, φ_{rf} is the phase shift due to the RF drive voltage relative to the RF voltage required for π is the phase shift, and $\varphi_{\Delta\lambda}$ is an additional phase term to account for the wavelength dependence of the dispersion. Note the necessary condition that the loss in the MZM and the fiber, the photodiode responsivity, and the input power all be the same for both wavelengths. In order for the cancelation to work, the proper delay must also be set at the end of the fiber before subtracting the two photocurrents such that $\varphi_{\Delta\lambda}$ can be set to zero. With this condition met, the subtraction of the photocurrents yields the following:

$$I_{\lambda 1} - I_{\lambda 2} = 2I_{DC}\varphi_{rf}\sin(\Omega t - \theta_1)\cos(\frac{1}{2}\theta_2) \tag{3.7c}$$

Looking at (3.7c), we can see two results: First, the second harmonic term has been canceled, and second, the fundamental term is doubled.

3.5.4 Single-Sideband Modulation

While the MZM can be used to cancel the second harmonic distortion introduced by dispersion, other modulation formats can be used to overcome dispersion in optical fibers. Single-sideband (SSB) modulation has been identified as a means to overcome the RF fading penalties associated with dispersion [51]. SSB modulation places the RF sideband on only one side of the carrier, rather than the more common dual-sideband modulation. Multiple methods have been demonstrated to generate SSB optical signals, including optical filtering of dual-sideband modulation [52], using a 90° RF hybrid with a single push-pull MZM [53] and using a single electrode MZM with bidirectional optical inputs [54]. Despite the advantages of using SSB modulation, the use of SSB is limited in RF photonic applications. One of the common misconceptions is that the RF output power of SSB modulation will be less than DSB modulation, as there is only one less sideband to mix with the optical carrier. However, work has been done to show that the SSB does not degrade the RF performance when compared to other types of DSB modulation from an MZM [55]. Specifically a dual-drive MZM using a 90° hybrid or a 180° hybrid or with the RF signal connected to just one arm of the MZM are compared. The various RF gains for each type of configuration are given in (3.8a)–(3.8d) using the small signal approximation of $J_n(\varphi_{rf}) \approx \varphi_{rf}^n / (2^n n!)$, with φ_{rf} as the phase shift in the MZM arms due to the RF signal. In (3.8), the R is the responsivity of the photodetector, α_{MZM} is the insertion loss of the MZM, P_{laser} is the power of the laser, and $\Delta\varphi_{dc}$ is the difference between the DC bias points for the two electrodes in the MZM. From this analysis, the SSB modulation from a 90° hybrid MZM has the same OIP3 performance as the three other cases. The SSB link does provide 3 dB less RF output power when compared to a DSB link using a 180° hybrid, but the SSB link has the same RF output power as the push-pull MZM and single-arm MZM. The SSB modulation increases the optical power in the sideband that is not nulled, making up for the loss of the other optical sideband.

$$P_{RF,Fund,ss} \approx (R\alpha_{MZM}P_{laser})^2 \sin^2(\Delta\varphi_{dc})\frac{\varphi_{rf}^2}{8}Z_{out}, \tag{3.8a}$$

$$P_{RF,Fund,ss} \approx (R\alpha_{MZM}P_{laser})^2 \sin^2(\Delta\varphi_{dc})\frac{\varphi_{rf}^2}{4}Z_{out} \tag{3.8b}$$

$$P_{RF,Fund,ss} \approx (R\alpha_{MZM}P_{laser})^2 \sin^2(\varphi_{dc1})\frac{\varphi_{rf}^2}{8}Z_{out}. \tag{3.8c}$$

$$P_{RF,Fund,ss} \approx (R\alpha_{MZM}P_{laser})^2 \sin^2(\varphi_{dc1})\frac{\varphi_{rf}^2}{8}Z_{out} \tag{3.8d}$$

3.5.5 Single-Sideband Modulation to Cancel Photodetector Nonlinearities

As mentioned previously, the second-order nonlinearity from the photodetector can become a limiting factor. In-depth studies have shown that the nonlinear photodetector response will be limited by the second-order distortion [17]. While specialty photodetectors have been demonstrated with improved nonlinear performance, it is desirable to improve the performance of commercially available photodiodes. One method uses a dual-output MZM at quadrature bias with balanced detection to cancel the photodetector-induced second-order nonlinearity in the HF band [25]. Phase-matching two parallel links operating in the multigigahertz range and over long fiber lengths becomes a significant challenge. Another method incorporates two MZMs inside a third overall MZM structure known as the dual-parallel MZM (DP-MZM). This type of modulator allows for the generation of SSB-modulated signals. Using two discrete MZMs has been investigated to linearize unwanted nonlinear distortions due to the individual MZMs [56]. Recent work has also demonstrated high-performance links using an integrated DP-MZM [57–59]. A recent demonstration uses the SSB modulation from a DP-MZM to cancel the unwanted second-harmonic distortion from the photodetector. Using the small-signal approximation of $J_n(\varphi_{rf}) \approx \varphi_{rf}^n / (2^n n!)$, with φ_{rf} as the phase shift in the MZM arms due to the RF signal, (3.9) shows that the generated photocurrent from the DP-MZM at the photodetector can be written as

$$
\begin{aligned}
I_{DC} &\approx 2\zeta \begin{bmatrix} 2+\cos(\varphi_{dc3}) - \cos(\varphi_{dc1}) - \cos(\varphi_{dc2}) - \cos(\varphi_{dc1}-\varphi_{dc3}) \\ -\cos(\varphi_{dc2}+\varphi_{dc3}) + \cos(\varphi_{dc1}-\varphi_{dc2}-\varphi_{dc3}) \end{bmatrix}, \\
I_{Fund}(t) &\approx 2\varphi_{rf}\zeta \begin{bmatrix} (\sin(\varphi_{dc1}) + \sin(\varphi_{dc1}-\varphi_{dc3}))\sin(\Omega_{rf}t) \\ +(\sin(\varphi_{dc2}) + \sin(\varphi_{dc2}+\varphi_{dc3}))\cos(\Omega_{rf}t) \\ +\sqrt{2}\sin(\varphi_{dc1}-\varphi_{dc2}-\varphi_{dc3})\sin(\Omega_{rf}t+\phi_{rf}) \end{bmatrix}, \\
I_{Second}(t) &\approx \frac{1}{2}\varphi_{rf}{}^2\zeta \begin{bmatrix} -(\cos(\varphi_{dc1}) + \cos(\varphi_{dc1}-\varphi_{dc3}))\cos(2\Omega_{rf}t) \\ +(\cos(\varphi_{dc2}) + \cos(\varphi_{dc2}+\varphi_{dc3}))\cos(2\Omega_{rf}t) \\ +2\cos(\varphi_{dc1}-\varphi_{dc2}-\varphi_{dc3})\cos(2\Omega_{rf}t+2\phi_{rf}) \end{bmatrix},
\end{aligned}
\tag{3.9}
$$

where $\zeta = R\alpha_{MZM}P_{laser}G_0/16, G_o$ is the optical gain from the EDFA after the DP-MZM output to the photodetector, is the photodetector responsivity, and $\varphi_{rf} = 3\pi/4$. From (3.9), the phase between the fundamental and the second harmonic can be adjusted by simply changing $\varphi_{\mathrm{dc}1,2,3}$. In order to cancel the photodetector-induced second harmonic, the phase of the second harmonic photocurrent from the DP-MZM is set to be 180° with relation to the second harmonic photocurrent generated from the photodetector by the fundamental optical power, which is shown in (3.10).

$$I_{PD}(P_{opt,\ in}) = -a_2\left(\frac{\varphi_{rf}\zeta}{\Re}\sigma_{comb}\right)^2 2\cos(2\Omega_{rf}t + 2\phi_{comb} + 2\phi_{dc}) \qquad (3.10)$$

The system uses the various bias control points of the DP-MZM to find the correct second harmonic. An experimental demonstration showed an improvement in the $\mathrm{OIP2}_{2\mathrm{H}}$ of 55.3 dBm. In addition, the third-order nonlinear distortion is shown to remain the same. However, the cancelation of the second harmonic comes at the cost of a reduction in the fundamental power of 6.3 dB.

3.6 Conclusions

Various advanced techniques have been demonstrated to improve the performance of the analog delay line. Various photonic components can be used to improve the overall RF metrics of the delay line. The photodiode nonlinearity can be improved with better component design. However, the nonlinearities also can be overcome by using different modulation formats. In addition, the limitations of the optical fibers can be overcome with the use of different fiber types and isolators. The noise of the EDFA can be characterized and properly controlled by properly designing the analog delay line to minimize the added noise from the noise penalty of each EDFA. Finally, the MZM can be used at different bias points to improve the RF performance. Chapter 4 develops the use of RF photonics for oscillators in the demonstration architecture.

References

[1] Funk, E. E., V. J. Urick, and F. Bucholtz, "High Dynamic Range, Long Haul (>100 km) Radio over Fiber," in *Microwave Photonics* (C. H. Lee, (ed.), Boca Raton, FL: CRC Press, 2007, pp. 185–212.

[2] Ackerman, E. I., et al., "RF-over-Fiber Links with Very Low Noise Figure," *Journal of Lightwave Technology,* Vol. 26, No. 15, 2008, pp.2441–2448.

[3] Karim, A., and J. Devenport, "Noise Figure Reduction in Externally Modulated Analog Fiber-Optic Links," *Photonics Technology Letters,* Vol. 19, No. 5, 2007, pp. 312–314.

[4] Joshi, A., S. Datta, and D. Becker, "GRIN Lens Coupled Top-Illuminated Highly Linear InGaAs Photodiodes," *Photonics Technology Letters,* Vol. 20, No. 17, 2008, pp. 1500–1502.

[5] Cox, C. H., et al. "Limits on the Performance of RF-over-Fiber Links and Their Impact on Device Design," *Transactions on Microwave Theory and Techniques,* Vol. 54, No. 2, 2006, pp. 906–920.

[6] Williams, K. J., and R. D. Esman, "Design Considerations for High-Current Photodetectors," *Journal of Lightwave Technology,* Vol. 17, No. 8, 1999, pp. 1443–1454.

[7] Li, Z., et al., "High-Saturation-Current Modified Uni-Traveling-Carrier Photodiode with Cliff Layer," *Journal of Quantum Electronics,* Vol. 46, No. 5, 2010, pp. 626–632.

[8] Lin, L. Y., et al., "High-Power High-Speed Photodetectors-Design, Analysis, and Experimental Demonstration," *Transactions on Microwave Theory and Techniques,* Vol. 45, No. 8, 1997, pp. 1320–1331.

[9] Urick, V. J., et al., "Fiber-Optic Links with All-Photonic RF Gain and Low RF Noise Figure," *International Microwave Symposium Digest,* Baltimore, MD, 2011.

[10] Williams, K. J., R. D. Esman, and M. Dagenais, "Nonlinearities in p-i-n Microwave Photodetectors," *Journal of Lightwave Technology,* Vol. 14, No. 1, 1996, pp. 84–96.

[11] Taylor, J., et al., "Characterization of Power-to-Phase Conversion in High-Speed P-I-N Photodiodes," *Photonics Journal,* Vol. 3, No. 1, 2011, pp. 140–151.

[12] Tulchinsky, D. A., and K. J. Williams, "Excess Amplitude and Excess Phase Noise of RF Photodiodes Operated in Compression," *Photonics Technology Letters,* Vol. 17, No. 3, 2005, pp. 654–656.

[13] Eliyahu, D., D. Seidel, and L. Maleki, "RF Amplitude and Phase-Noise Reduction of an Optical Link and an Opto-Electronic Oscillator," *Transactions on Microwave Theory and Techniques,* Vol. 56, No. 2, 2008, pp. 449–456.

[14] Devgan, P., "Review of Improvements in Radio Frequency Photonics," *AFRL Report,* AFRL-RY-WP-TR-2017-0156, 2017.

[15] Hastings, A. S., D. A. Tulchinsky, and K. J. Williams, "Photodetector Nonlinearities Due to Voltage-Dependent Responsivity," *Photonics Technology Letters,* Vol. 21, No. 21, 2009, pp. 1642–1644.

[16] McKinney, J. D., et al., "Measurement of Photodiode Harmonic Distortion Using Optical Comb Sources And High-Resolution Optical Filtering," *Conference on Lasers and Electro-Optics,* 2009.

[17] Urick, V. J., et al., "Wideband Analog Photonic Links: Some Performance Limits and Considerations for Multioctave Implementations," *International Society for Optics and Photonics OPTO,* 2012, pp. 825904–825904.

[18] Ackerman, E. I., "Broadband Linearization of a Mach-Zehnder Electrooptic Modulator," *Transactions on Microwave Theory and Techniques,* Vol. 47, No. 12, 1999, pp. 2271–2279.

[19] Ismail, T., et al., "High-Dynamic-Range Wireless-over-Fiber Link Using Feedforward Linearization," *Journal of Lightwave Technology,* Vol. 25, No. 11, 2007, pp. 3274–3282.

[20] O'Connor, S. R., T. R. Clark, Jr., and D. Novak, "Wideband Adaptive Feedforward Photonic Link," *Journal of Lightwave Technology,* Vol. 26, No. 15, 2008, pp.2810–2816.

[21] Urick, V. J., et al., "Phase Modulation with Interferometric Detection as an Alternative to Intensity Modulation with Direct Detection for Analog-Photonic Links," *Transactions on Microwave Theory and Techniques,* Vol. 55, No. 9, 2007, pp.1978–1985.

[22] Hastings, A., et al., "Suppression of Even-Order Photodiode Nonlinearities in Multioctave Photonic Links," *Journal of Lightwave Technology,* Vol. 26, No. 15, 2008, pp. 2557–2562.

[23] Urick, V. J., et al., "Photodiode Linearity Requirements for Radio-Frequency Photonics and Demonstration of Increased Performance Using Photodiode Arrays," *Proc. Topical Meeting Microwave Photonics,* Gold Coast, Australia, 2008, pp.86–89.

[24] Roman, J. E., et al., "Optical Fiber Cables for Synchronous Remoting of Numerous Transmitters/Receivers," *Photonics Technology Letters,* Vol. 10, No. 4, 1998, pp. 591–593.

[25] Urick, V. J., et al., "Field Test on the Feasibility of Remoting HF Antennas with Fiber Optics," *NRL Memorandum Report,* NRL/MR/5652-08-9137, 2008.

[26] Klamkin, J., et al., "Unitraveling-Carrier Balanced Photodiode with Tunable MMI Coupler for Optimization of Source Laser RIN Suppression," *Proc. Topical Meeting Microwave Photonics,* Valencia, Spain, 2009.

[27] Theys, M. D., et al., "Using a Microcontroller to Optimize the Bias Voltage of Balanced Photodiodes to Minimize Even-Order Distortion in Analog Fiber-Optic Links," *Proc. Optical Fiber Communication Conference,* 2010.

[28] Grüner-Nielsen, L., et al., "Dispersion Compensating Fibers," *Optical Fiber Technology,* Vol. 6, No. 2, 2000, pp. 164–180.

[29] Akasaka, Y., et al., "Dispersion Compensating Optical Fiber for Wavelength Multiplex Transmission and Method Using Same," United States Patent No. 5,568,583, 1996.

[30] Vengsarkar, A. and J. Wagener, "Dispersion Compensating Optical Fiber and Communication System Comprising Same," United States Patent No. 5,802,234, 1998.

[31] Onishi, M., and H. Kanamori, "Optical Transmission System with Dispersion Compensating Optical Fiber," United States Patent No. 5,555,340, 1996.

[32] Liu, A., "Suppressing Stimulated Brillouin Scattering in Fiber Amplifiers Using Nonuniform Fiber and Temperature Gradient," *Optics Express,* Vol 15, No. 3, 2007, pp. 977–984.

[33] Parker, T. R., et al., "Temperature and Strain Dependence of the Power Level and Frequency of Spontaneous Brillouin Scattering in Optical Fibers," *Optics Letters,* Vol. 22, No. 11, 1997, pp. 787–789.

[34] Dragic, P. D., et al., "Optical Fiber with an Acoustic Guiding Layer for Stimulated Brillouin Scattering Suppression," *Proc. Conference on Lasers and Electro-Optics,* 2005.

[35] Mao, X. P., et al., "Stimulated Brillouin Threshold Dependence on Fiber Type and Uniformity," *Photonics Technology Lett*ers, Vol. 4, No. 1, 1992, pp. 66–69.

[36] De Oliveira, C. A. S., et al. "Stimulated Brillouin Scattering in Cascaded Fibers of Different Brillouin Frequency Shifts," *Journal of Optical Society America B,* Vol. 10, No. 6, 1993, pp. 969–972.

[37] Devgan, P. S., et al., "SOA-Based Regenerative Amplification of Phase Noise Degraded DPSK Signals," *Proc. Optical Fiber Communication Conference,* 2005.

[38] Devgan, P. S., et al., "Highly Efficient Multichannel Wavelength Conversion of DPSK Signals," *Journal of Lightwave Technology,* Vol. 24, No. 10, 2006, pp.3677–3682.

[39] Becker, P. C., N. A. Olsson, and J. R. Simpson, *Erbium-Doped Fiber Amplifiers Fundamentals and Technology,* San Diego, CA: Academic Press, 1999.

[40] Haus, H. A., "The Noise Figure of Optical Amplifiers," *Photonics Technology Letters,* Vol. 10, No. 11, 1998, pp. 1602–1604.

[41] Urick, V. J., F. Bucholtz, and K. J. Williams, "Noise Penalty of Highly Saturated Erbium-Doped Fiber Amplifiers in Analog Links," *Photonics Technology Letters,* Vol. 18, No. 6, 2006, pp. 749–751.

[42] Devgan, P. S., et al., "Cascaded Noise Penalty For Amplified Long-Haul Analog Fiber-Optic Links," *Transactions on Microwave Theory and Techniques,* Vol. 55, No. 9, 2007, pp. 1973–1977.

[43] Urick, V. J., et al., "The Performance of Analog Photonic Links Employing Highly Compressed Erbium-Doped Fiber Amplifiers," *Transactions on Microwave Theory and Techniques,* Vol. 54, No. 7, 2006, pp. 3141–3145.

[44] Wooten, E. L., et al. "A Review of Lithium Niobate Modulators for Fiber-Optic Communications Systems," *Journal of Selected Topics in Quantum Electronics,* Vol. 6, No. 1, 2000, pp. 69–82.

[45] Urick, V. J., et al.," Analysis of an Analog Fiber-Optic Link Employing a Low-Biased Mach–Zehnder Modulator Followed by an Erbium-Doped Fiber Amplifier," *Journal of Lightwave Technology,* Vol. 27, No. 12, 2009, pp. 2013–2019.

[46] Foreman, S. M., et al. "Remote Transfer of Ultrastable Frequency References Via Fiber Networks," *Review of Scientific Instruments,* Vol. 78, No. 2, 2007, 021101.

[47] Devgan, P. S., et al. "Even-Order Harmonic Cancellation for Off-Quadrature Biased MZM with Improved RF Metrics Using Dual Wavelength Inputs and Dual Outputs," *Optics Express,* Vol. 17, No. 11, 2009, pp. 9028–9039.

[48] Rollins, D., "Linearized Optical Link Using a Single MZM and Two Optical Carriers," United States Patent No. 7,079,780, 2006.

[49] Schmuck, H., "Comparison of Optical Millimetre-Wave System Concepts with Regard to Chromatic Dispersion," *Electronics Letters,* Vol. 31, No. 21, 1995, pp. 1848–1848.

[50] Corral, J. L., J. Marti, and J. M. Fuster, "General Expressions for IM/DD Dispersive Analog Optical Links with External Modulation or Optical Up-Conversion in a Mach-Zehnder Electrooptical Modulator," *Transactions on Microwave Theory and Techniques,* Vol. 49, No.10, 2001, pp. 1968–1976.

[51] Hraimel, B., et al., "Optical Single-Sideband Modulation with Tunable Optical Carrier to Sideband Ratio in Radio Over Fiber Systems," *Journal of Lightwave Technology,* Vol. 29, No. 5, 2011, pp. 775–781.

[52] Park, J., W. V. Sorin, and K. Y. Lau, "Elimination of the Fibre Chromatic Dispersion Penalty on 1550 nm millimetre-Wave Optical Transmission," *Electronics Letters,* Vol. 33, No. 6, 1997, pp. 512–513.

[53] Smith, G. H., D. Novak, and Z. Ahmed, "Technique for Optical SSB Generation to Overcome Dispersion Penalties in Fibre-Radio Systems," *Electronics Letters,* Vol. 33, No. 1, 1997, pp. 74–75.

[54] Loayssa, A., D. Benito, and M. J. Garde, "Single-Sideband Suppressed-Carrier Modulation Using a Single-Electrode Electrooptic Modulator," *Photonics Technology Letters,* Vol. 13, No. 8, 2001, pp. 869–871.

[55] Devgan, P. S., et al., "Cancellation of Photodiode-Induced Second Harmonic Distortion Using Single Side Band Modulation from a Dual Parallel Mach-Zehnder," *Optics Express,* Vol. 20, No. 24, 2012, pp. 27163–27173.

[56] Korotky, S. K., and R. M. de Ridder, "Dual Parallel Modulation Schemes for Low-Distortion Analog Optical Transmission," *Journal on Selected Areas in Communications,* Vol. 8, No. 7, 1990, pp. 1377–1381.

[57] Zhu, G., W. Liu, and H. R. Fetterman, "A Broadband Linearized Coherent Analog Fiber-Optic Link Employing Dual Parallel Mach–Zehnder Modulators," *Photonics Technology Letters,* Vol. 21, No. 21, 2009, pp. 1627–1629.

[58] Li, S., et al., "Highly Linear Radio-Over-Fiber System Incorporating a Single-Drive Dual-Parallel Mach–Zehnder Modulator," *Photonics Technology Letters,* Vol. 22, No. 24, 2010, pp. 1775–1777.

[59] Kawanishi, T., and M. Izutsu, "Linear Single-Sideband Modulation for High-SNR Wavelength Conversion," *Photonics Technology Letters,* Vol. 16, No. 6, 2004, pp. 1534–1536.

4

Oscillators Utilizing RF Photonics

Returning to the demonstration architecture for using RF photonics, the emphasis of this chapter is on oscillators. The oscillator is used for many different applications. The oscillator offers a common clock signal to all the other subsystems in the architecture. It can be used for downconverting the RF signal or for triggering a sampling function, such as an ADC. The oscillator plays a significant role in the RF architecture and can be implemented using RF photonic techniques. This chapter discusses the need for high-quality and fast-tuning oscillators. In addition, the chapter reviews various techniques for making oscillators based on RF photonic technologies, with an emphasis on the optoelectronic oscillator and the beating of two laser sources to generate an RF tone.

4.1 Need for Oscillators

One of the common signal-processing requirements many varied applications share is the need for very precise timing. Doppler radars require very low phase noise clock sources in order to minimize the uncertainty of objects that are being tracked [1]. As optical digital networks have been shown to transmit an aggregate of 10 Tbps of data [2], they require low phase noise clocks for multiplexing the data streams on the entire network. Other applications requiring low phase noise clocks include sampling for ADCs [3, 4], clock recovery [5], and pulse sources [6]. In computers, on-chip optical interconnects are being investigated to overcome the clock skew and speed limitations that copper interconnects currently exhibit [7, 8]. In addition, as the processing speeds of the chips increase beyond 1 GHz, the requirement on the phase noise of the data clock becomes more stringent. Thus, a photonic low phase noise clock that synchronizes the data bus on the chip becomes very important. On the opposite end, optical networks are used to send voice, video, and internet data of 10 Gbps or higher over thousands of kilometers [9]. The ability to synchronize these signals for proper routing and processing is essential to meet the ever-increasing demand for bandwidth. Again the low phase noise photonic clock becomes a key part of the network. A final example of the importance of low phase noise photonic clocks is in their use in ADCs. The sample and hold circuit of the ADC depends on a very low jitter clock to operate correctly. Any error in the timing of the clock can lead to degraded SNR for the ADC [10], where ENOB is the effective number of bits.

$$SNR(dB) = 20\log_{10}\left(\frac{1}{2\pi f_{rf} t_{jitter}}\right) \tag{4.1a}$$

$$SNR(dB) = 6.02 * ENOB + 1.78 \tag{4.1b}$$

Equation (4.1a) defines the dependence of the SNR of the ADC as a function of both the timing jitter of the sampling clock (t_{jitter}) and the frequency of the input sinusoidal signal to the ADC (f_{rf}). Equation (4.1b) then shows the relation of SNR to the number of bits in the ADC [11]. Then the following relationship can be derived $t_{jitter} \approx 1/\left(f_{samp} 2^{N_{bits}+2}\right)$, where f_{samp} = $2^* f_{\text{rf}}$ is the Nyquist sampling frequency of the ADC. Plotting this relationship in Figure. 4.1 shows the required timing jitter for a given ENOB and sampling rate of the ADC. As the ADCs have increased to sampling rates of 1 GHz and higher, the requirement to get even 10 ENOB forces the timing jitter to be on the order of tens of femtoseconds. ADCs place the most stringent requirements on the timing jitter of the clock and are one of the strongest drivers for low phase noise sources.

4.2 Phase Noise and Timing Jitter

Before discussing examples of photonic-based oscillators, a brief discussion of phase noise and its relationship to timing jitter is required. In characterizing oscillators of any type, the common measurement that is made of the frequency fluctuations is called the phase noise measurement. At least two formal definitions of phase noise have been presented in the literature, which

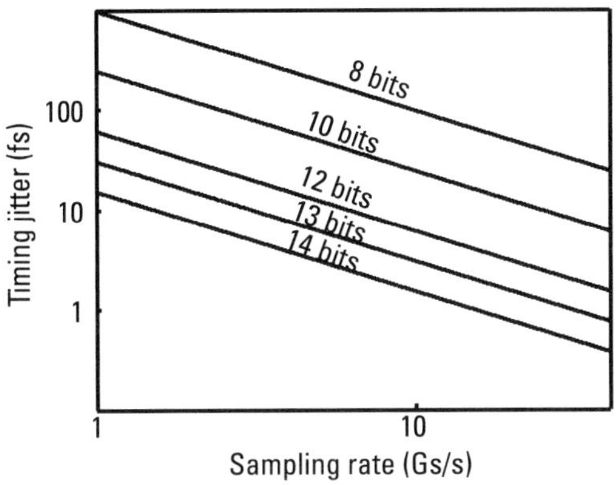

Figure 4.1 Required timing jitter for an ADC at a sampling rate between 1 and 40 Gbps for 8, 10, 12, 13, and 14 ENOB resolution. (After: [12].)

can lead to some confusion [13]. In this discussion, the accepted definition, denoted as $L(f)$, is the single-sideband power spectral density (PSD). The PSD is normalized to the power of the RF signal and has units of decibels relative to carrier power per one hertz bandwidth.

In general, the PSD will have contributions from both amplitude and phase noise. However, the measurement system can be configured to separate the amplitude noise and the phase noise contributions to the PSD. One example uses a phase detector, which converts any phase fluctuations between two input signals into a voltage that can be measured [14]. Other methods also exist to measure only the phase noise contribution to the PSD. In all of these cases, the amplitude noise does not contribute to the measured PSD. In this case, the timing jitter can be extracted directly from the measured PSD using thc following relation

$$\sigma_n = \sqrt{2\int_{f_{\min}}^{f_{\max}} \mathcal{L}(f)\, df} \tag{4.2}$$

where σ_n is the root-mean-squared (rms) total noise, and $f_{\max}$ and $f_{\min}$ are the maximum and minimum frequency offsets that the phase noise is integrated over. Since the rms amplitude noise (σ_a) contribution is negligible, the rms timing jitter (σ_j) can be obtained by the following relationship $\sigma_j = \sigma_n/(2\pi f_{rf})$ where f_{rf} is the RF frequency of the clock. However, if the measurement system is not set to ignore the amplitude contribution, further work is needed to extract the timing jitter contribution from the rms total noise. When the higher harmonics of the RF signal can be measured, it has been shown previously [15] that the rms amplitude noise and rms timing jitter components can be separated from the overall rms noise by the following relationship $\sigma_n^2 = n^2 \left(2\pi f_{rf}\right)^2 \sigma_j^2 + \sigma_a^2$ where n is the n-th harmonic of the RF signal being measured. The noise PSD of both the timing jitter and the amplitude noise is plotted in Figure 4.2. Note that the amplitude noise spectra is arbitrarily plotted broader than the timing jitter noise in order to differentiate the two sources of noise and is not necessarily representative of a real system. As the noise of each harmonic is measured, it will increase quadratically with the timing jitter noise while the amplitude noise contribution is constant. Thus, if one can measure the noise of a sufficient number of harmonics and then fit the results to a curve, the timing jitter and amplitude noise contributions can be determined.

4.3 Optoelectronic Oscillator

One of the most used examples of RF photonics for oscillators is the opto-elect (OEO). The OEO was first proposed by Yao and Maleki [16–18] in 1996. The OEO is similar to optoelectronic feedback circuits demonstrated by Neyer and Voges [19] in 1982, by Nakazawa, et al. [20] in 1984, and by Lewis [21] in 1992.

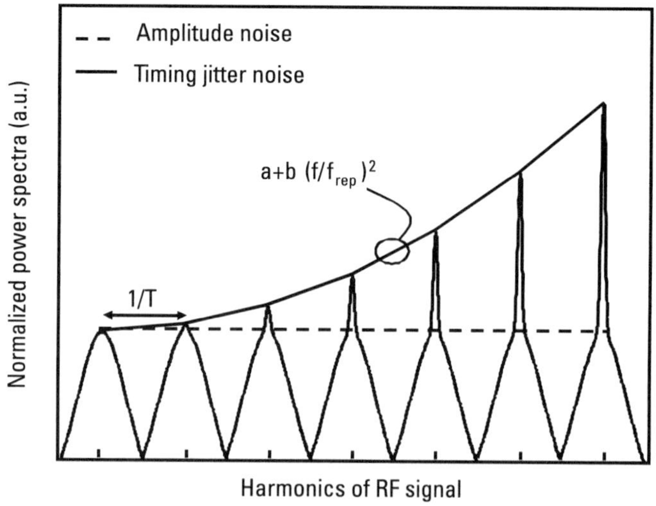

Figure 4.2 Theoretical noise PSD for multiple harmonics of the RF frequency showing the timing jitter and amplitude noise contributions. (After: [12].)

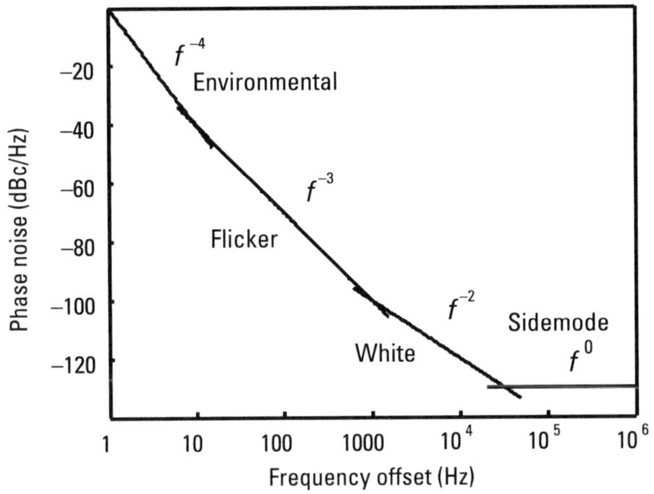

Figure 4.3 Diagram of a basic OEO (where PD stands for photodiode and opt. mod. stands for optical modulator). (After: [12].)

As seen in Figure 4.3, the basic OEO starts with a continuous wave (CW) laser feeding an optical modulator. The optical output of the modulator is passed through a length of optical fiber into a photodiode. The recovered electrical signal is amplified and sent through an electronic bandpass filter. The output of the filter is connected to the RF input of the modulator in order to complete the optoelectronic cavity. When the gain within the cavity is greater than the loss, the OEO will begin to oscillate. The electronic bandpass filter selects the frequency of oscillation by attenuating the other free-running modes of the cavity.

The OEO differs from the previous optoelectronic circuits by including a very low loss fiber-optical delay, which is essential to creating a cavity with a very high Q factor. In the case of the OEO, The Q factor is defined as the ratio of the stored energy in the cavity over the loss of the cavity. With optical fibers having a loss of 0.2 dB/km, the cavity can store a large amount of energy with very little loss. Thus, the Q factor of the OEO can easily achieve the level of 10^8 or higher. This translates to a 10-GHz clock signal with a phase noise of −140 dBc/Hz at 10 kHz offset [23]. Unfortunately the phase noise of the OEO cannot be continuously improved as the length of fiber increases. Lasri et al. [22] demonstrated the improvement in the timing jitter of an OEO having an inverse square root dependence on the fiber length. The limitations in that case were due to the random fiber length variations due to small fluctuations in the environment around the fiber span.

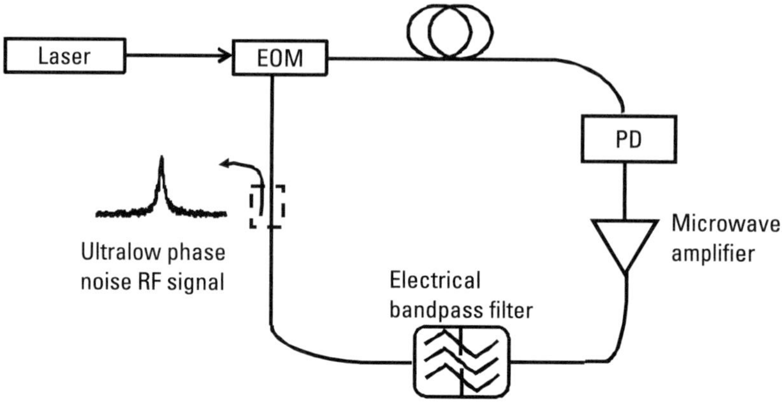

Figure 4.4 Theoretical phase noise plot showing the different influences of noise on the PSD at frequency offsets from 1 Hz to 1 MHz from the carrier. (After: [12].)

To show the general shape of the PSD, a typical theoretical phase noise plot of an OEO is presented in Figure 4.4. The oscillator's performance is determined by different sources of noise at different frequency offsets [13]. In the 1–10-Hz frequency offset range, the noise is dominated by environmental fluctuations, caused by temperature variations and acoustic vibrations in the fiber-based cavity, and has a slope of f^{-4}. In the 10-Hz–1-kHz frequency offset range, the noise is dominated by flicker phase noise from the RF amplification stage necessary for most photonics-based oscillators to oscillate and has a slope of f^{-3}. The next frequency range of 1 kHz–30 kHz is dominated by white phase noise and has a slope of f^{-2}. Finally, at frequency offsets of 30 kHz and higher, the phase noise has a flat noise response that is due to the adjacent nonoscillating side mode of the OEO.

One of the advantages of the OEO is the high performance achieved with the use of commercially available parts that are inexpensive. Once the first OEO was demonstrated, multiple demonstrations were presented to improve on the initial design. The following sections present some examples of improved OEOs.

4.3.1 Multiloop OEO

One of the disadvantages of the initial OEO design was the limitation of the electrical bandpass filter to suppress unwanted side modes. The fiber length must be maximized in order to achieve as high a Q factor as possible. As the fiber length increases, the spacing between the cavity modes decreases. For example, a 3-km length of fiber will yield a cavity mode spacing of approximately 67 kHz [24]. A high-quality electrical bandpass filter at 10 GHz has a

3-dB bandwidth of 10 MHz [23]. Thus, many side modes will pass through the electrical bandpass filter and adversely affect the phase noise. One method to reduce this problem was to introduce a second fiber length into the OEO cavity [25]. An example of this type of OEO is shown in Figure 4.5. The second loop of the OEO has its own set of cavity modes. As long as the length of the second loop is not a harmonic multiple of the first loop, the cavity modes will not overlap with each other, as seen below the OEO in Figure 4.5. However, the modes from each loop that are closest to each other will lock and suppress the other cavity modes. This can be seen in Figure 4.6, which shows the single-loop phase noise spectrum with the side modes along with the dual-loop spectrum with the side modes suppressed below –110 dBc/Hz. The trade-off of the system is that the phase noise is now an average of noise of the two loops individually, not the phase noise of just the longer loop. In addition, since both loops support the side modes, they are not completely

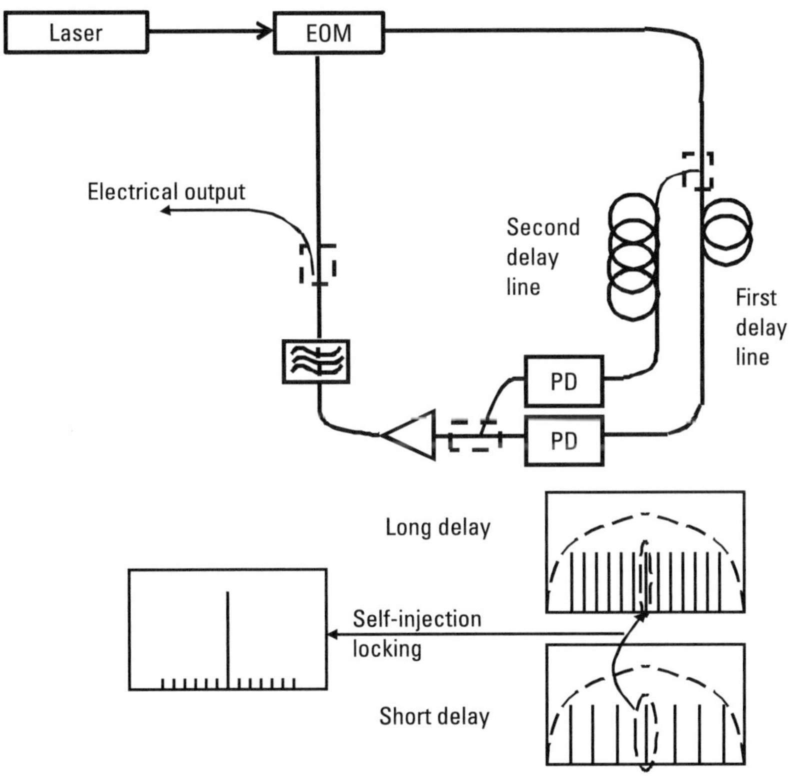

Figure 4.5 Diagram of a dual-loop OEO along with the cavity modes within the filter passband for the two fiber delays. (PD: photodiode; mod: optical modulator; and EBF: bandpass filter.) (After: [12].)

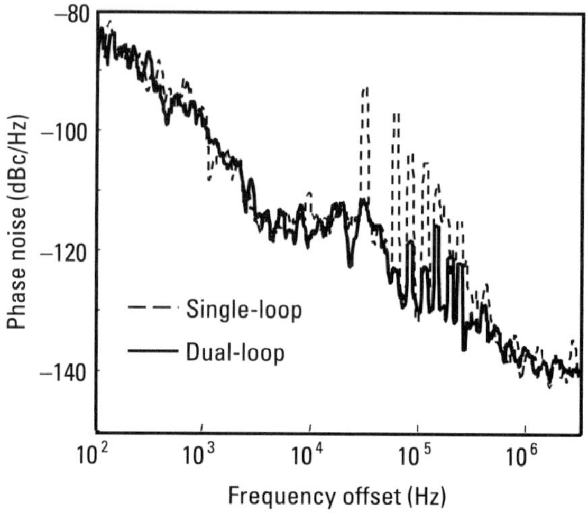

Figure 4.6 Measured phase noise of the single-loop and dual-loop OEO showing the non-oscillating side mode suppression below -110 dBc/Hz. (After: [12].)

eliminated, but merely suppressed. Nevertheless, many demonstrations have been made with the multiloop OEO in order to suppress the side modes [26–28].

In order to address the shortcomings of the multiloop OEO, a dual-cavity OEO was conceived and demonstrated by Zhou and Blasche [29], in which a master-slave configuration was used. In this demonstration, a short-loop OEO, known as the slave, is injection-locked by a longer loop OEO, known as the master. An example of the master-slave OEO appears in Figure 4.7. Equivalent systems using electronic oscillators have shown that the slave oscillator will have the phase noise performance of the master oscillator. In this case, the overall system has the widely spaced cavity modes of the

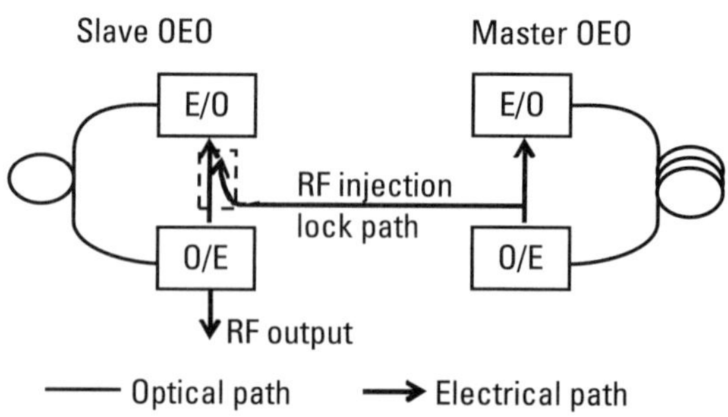

Figure 4.7 Diagram of a master-slave OEO. (DFB: distributed feedback laser; PD: photodiode; MZM: Mach Zehnder modulator; and EBF: electrical bandpass filter.)

short-loop OEO while having the phase noise performance of the longer loop OEO. The demonstrated phase noise of the generated 10-GHz RF signal is –130 dBc/Hz at an offset of 1 kHz [29]. The nonoscillating side modes are less than –140 dBc/Hz, which is lower than the side modes in the dual-loop configurations shown previously [25]. This configuration avoids the averaging of the two fiber loop's phase noise, with the added complexity of having two entirely separate OEO cavities. Further work has been done to analyze this system more closely and further demonstrations have been made [30, 31].

4.3.2 OEO with All-Photonic Gain

One of the limiting noise sources for the OEO is the RF amplifier necessary for the OEO to oscillate. Even though the loss of the optical fiber in the cavity is very low, both the RF-to-optical conversion at the optical modulator and the optical to RF conversion at the photodetector have loss. Therefore the electronic RF amplifier is necessary to overcome both of these losses. However, like any amplifier, it adds additional phase noise to the OEO, as shown in Figure 4.4. While the amplifier is used in the OEO, it will become compressed as it provides unity gain for the cavity to meet the oscillation condition. In compression, amplitude fluctuations will become suppressed, but the phase fluctuations will increase. This leads to the added phase noise. While a lot of work has been done to reduce the phase noise introduced by the electronic RF amplifiers used in OEOs, they are expensive and not wideband. Therefore, each OEO would need its own special RF amplifier to operate at a given frequency, limiting the otherwise potentially wideband operation of an OEO. If the OEO can oscillate utilizing only photonic gain, the wideband operation can be preserved. One method for providing all-photonic gain is the low-biased MZM followed by an optical amplifier. As discussed in Chapter 3, the low-biased MZM can provide increased RF gain without the need of an electronic amplifier. A 10-GHz OEO using all-photonic gain has been previously demonstrated [32, 33]. In this demonstration, the RF gain has to match the loss in the OEO cavity, which is 13 dB. Looking at Figure 4.8, this would require a photocurrent of approximately 60 mA, which is well above the power-handling capabilities of the photodetector. So a different method is required to meet the oscillation condition. By using an optical amplifier after the MZM and lowering the DC bias of the MZM, the RF gain of the OEO can be increased to a level that matches the loss of the OEO cavity. One advantage of the all-photonic OEO is that the RF power is increased while the photocurrent is kept at 6 mA, a level that the photodetector can handle. Looking at Figure 4.8, the RF gain at 6 mA for the all-photonic gain is 21.5

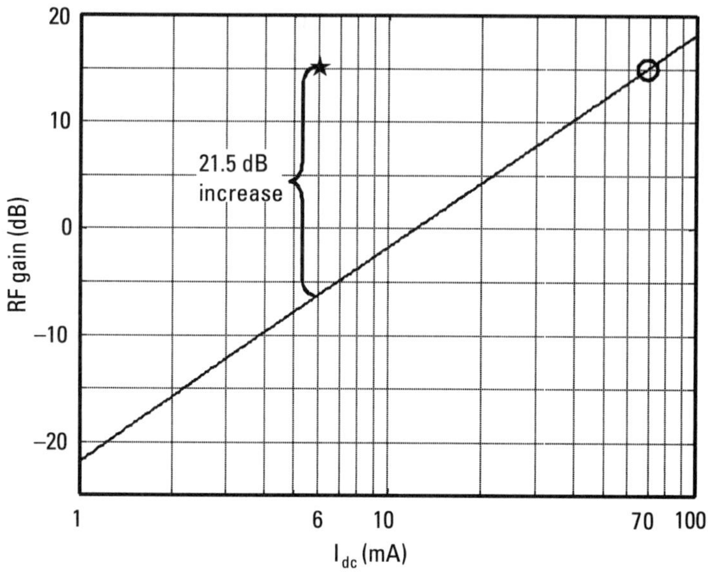

Figure 4.8 Theoretical plot of the RF gain for a single pass in the OEO as a function of DC photocurrent. (After: [12].)

dB higher than in the quadrature bias case. A disadvantage of the all-photonic gain OEO is that the second harmonic of the RF signal becomes quite large. However, in the case of the OEO, the electrical bandpass filter will suppress the second harmonic and make it negligible. The all-photonic gain OEO also showed an improvement in the phase noise as compared to the same OEO used with an electrical RF amplifier. The phase noise spectrum shows that the noise is no longer dominated by the flicker phase noise in the 100 Hz–1 kHz frequency offset range. In addition, the overall phase noise spectrum of the all-photonic OEO is lower than in the electronic RF amplifier OEO, in part due to the lower RF noise figure of the overall system. Thus, the all-photonic gain OEO has several advantages over electronic RF amplifier OEOs.

4.3.3 Clock Synchronization Using an OEO

The OEO can also be used for clock synchronization. As discussed previously, the outputs of two oscillators can be synchronized by injection locking. The theory of injection locking is a well-studied subject—after being first realized by Huygens, who discovered in the late 1800s that the pendulum clocks on the wall of his bedroom would become synchronized if they were placed within a certain distance of each other [34]. The locking mechanism occurred through mechanical vibrations transmitted through the wall. As seen in the movie at the website found on page *iv* of this book, the two pendulums are out of sync. As the mechanical vibrations increase, the pendulums become synchronous. Since then, injection locking has been observed in many different systems. In fact, the injection locking process is what allows the multiloop OEO to combine the RF signal between the different loops to yield a single low-noise RF signal. In order to use the OEO for clock recovery, the OEO can be injection-locked by an external signal. The output of the OEO will then become synchronized with the incoming RF signal. Before presenting examples of clock synchronization, we will review the basics of the injection locking process.

The injection locking process for electronic oscillators was first formally described by Adler [35]. One oscillator is termed as the injecting oscillator, also called the master oscillator, as it is the source of the signal, and the other oscillator is the free-running oscillator, or the slave oscillator, as it is the oscillator to be locked to the injecting oscillator. As long as the frequency of the RF signal is within the locking range of the slave oscillator, the slave oscillator will become both frequency- and phase-locked to the incoming RF signal. Adler developed mathematical expressions for the locking phenomena between the electronic oscillators and determined the relationship for locking as a function of frequency detuning between the two oscillators. While the initial work involved electronic oscillators, the theory can be expanded to any oscillator, including optical oscillators such as lasers.

As mentioned above, the OEO clock synchronization performance can be measured by looking at the phase noise of the OEO when it is injection-locked. Following a similar derivation as that shown in [36], the phase noise of the slave OEO can be described in relationship to the phase noise of the injected signal. To begin, the frequency-locking range is defined as $\omega_{\text{lock}} = 2\eta\omega_{BW}$, where $\eta = A_{inj}/A_{free}$, $\omega_{BW} = \omega_{free}/2Q$, A_{free} is the free-running OEO's signal amplitude, A_{inj} is the injected signal's amplitude, Q is the quality factor of the OEO, and $\omega_{\text{free}} = 2\pi f_{\text{free}}$ is the free-running OEO's angular frequency. These two components give some insight into the injection locking process of the OEO. First, the higher the Q of the OEO, the narrower the range of frequencies over which the OEO can be injection-locked. Second, the stronger the injected signal amplitude (or power) is compared to the free-running OEO, the wider the locking range is. The locking range can be adjusted by controlling these two parameters for use in a given application.

Now the phase noise transfer can be discussed. First, $\mathcal{L}_{\text{inj}}(\omega)$, $\mathcal{L}_{\text{free}}(\omega)$, and $\mathcal{L}_{OEO}(\omega)$ are defined as the single-sideband phase noise for the injected signal, the free-running slave OEO RF signal, and the injection-locked OEO oscillator, respectively. Then looking at the case where the signal is injected into the OEO cavity, the following can be written [37]

$$\mathcal{L}_{OEO}(\omega) = P_{transfer}(\omega)\mathcal{L}_{inj}(\omega) + P_{transfer}(\omega)\mathcal{L}_{free}(\omega)\left(\frac{2\omega}{\omega_{lock}}\right)^2$$
$$P_{transfer}(\omega) = \frac{1}{1+\left(\frac{2\omega}{\omega_{lock}}\right)^2} \tag{4.3}$$

where $P_{\text{transfer}}(\omega)$ represents the phase noise transfer function of the oscillator and has the form of a lowpass filter.

Figure 4.9 shows the calculated PSD of the phase noise, measured in units of decibels relative to carrier power per one hertz bandwidth, as a function of offset frequency from the carrier for an injection-locked OEO. The frequency-offset range is between 100 Hz and 10 MHz. The phase noise of the free-running OEO and the injected signal both follow an ideal $1/f^2$ relationship (dominated by white noise), with the free-running OEO phase noise approximately 20 dB lower than that of the injected signal. For three different locking ranges, the phase noise of the injection-locked OEO follows that of the injected signal within the various locking ranges. Outside the locking range, the phase noise decreases to the level of the free-running OEO phase noise, following the slope of $P_{\text{transfer}}(\omega)$. With this description in place, some different demonstrations of clock synchronization can be presented.

Using an OEO, clock synchronization was demonstrated by measuring the phase noise of the OEO before and after injecting the cavity with an RF signal [38]. The phase noise results are plotted in Figure 4.10. Like the theoretical plot in Figure 4.9, the phase noise of the injected signal and the

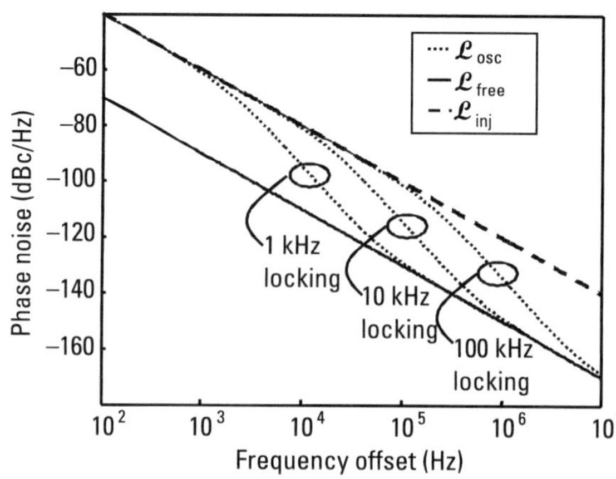

Figure 4.9 Theoretical plot of the phase noise of an injection-locked OEO along with the injected signal and free-running OEO's phase noise. (After: [12].)

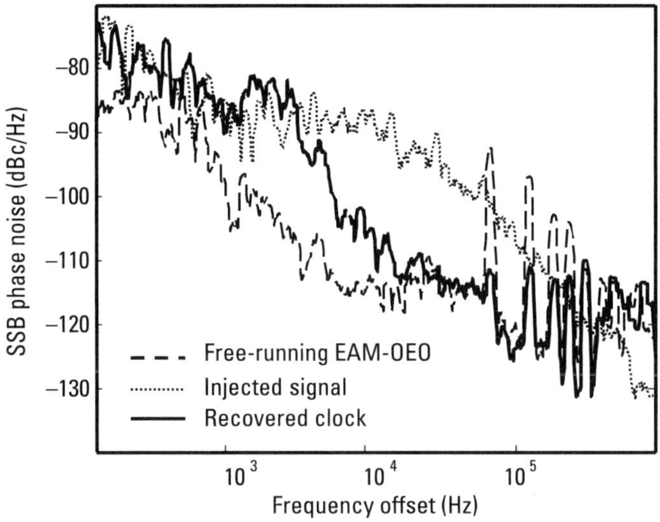

Figure 4.10 Measured plot of the phase noise of an injection-locked OEO along with the injected signal and free-running OEO's phase noise. (After: [12].)

free-running OEO are separate, and the phase noise of the injection-locked OEO follows the injected signal until about 1 kHz offset and then decreases to the phase noise of the free-running OEO. Note that the measured phase noise does decrease with a –20 dB/decade slope as predicted by the theory between the (–85 dBc at 1 kHz offset to –105 kHz at 10 kHz offset). In order to increase the locking range, the injected optical power into the OEO is increased by approximately 12 dB while keeping everything else the same. With the higher injected power, the locking range increases just less than 20 times, which again matches the theoretical prediction.

The clock synchronization can also be measured in the time and frequency domain. In this case, the RF output of the OEO is split to both the input of an electrical spectrum analyzer (for frequency domain measurements) and the trigger input of a high-speed sampling scope (for time domain measurements). A portion of the injected signal is connected to the RF input of the high-speed sampling scope. The frequency and time domain plots appear in Figure 4.11 for three different cases. In the first case, the injected signal's frequency is outside the locking range. The frequency measurement shows the free-running signal and the injected signal, along with beat notes between the two. The time measurement shows random noise as the trigger signal is not phase-locked to the input signal of the scope. The second case shows the free-running OEO and the injected signal almost phase-locked. Again the frequency domain measurement shows beat notes, and the time

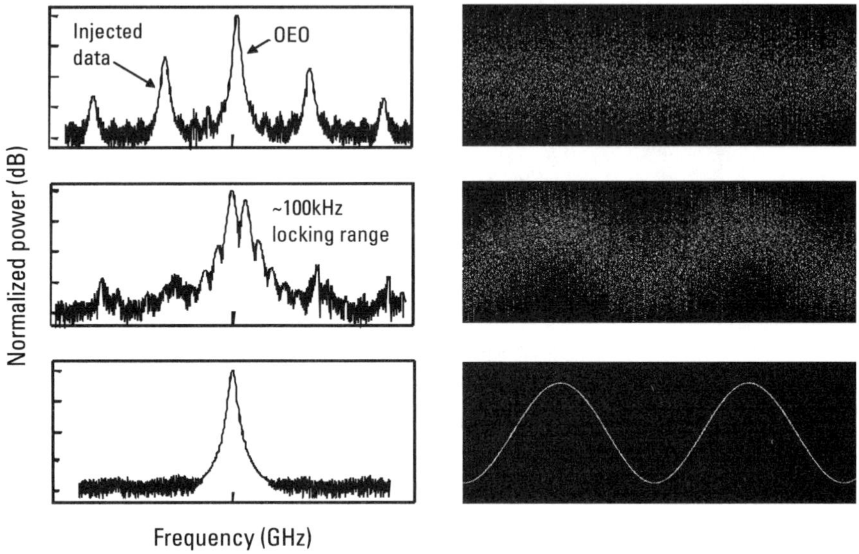

Figure 4.11 Measured plot of the frequency and time domain for the OEO when the injected signal is outside the locking range, just outside the locking range, and injection-locked. (After: [12].)

domain measurement is still noisy. The third case shows when the OEO and the injected signal are phase-locked. Now, the frequency measurement shows a single tone, and the time domain measurement clearly shows the injected sinusoidal signal. This demonstration shows that the OEO-generated signal is locked to the injected signal and can be used for synchronization purposes.

Other examples of injection-locked OEOs exist using different modulators, higher frequencies, and different injected signals [39–41]. Work continues to use these systems for clock synchronization for various applications.

While OEOs were originally used for clock signal generation and synchronization, they have also been investigated for many other signal-processing applications [12]. OEOs have very good phase noise performance. Unfortunately, the trade-off is the ability to be tunable. The very nature of the OEO limits its ability to be tuned very quickly in frequency. Some applications may require an oscillator that can be tuned very quickly.

4.4 Oscillators Based on Two Laser Sources

Beyond OEOs, other methods exist for generating clock signals using photonics. One of the most common methods involves the beat signal between two lasers. Starting with two laser sources that operate at two different optical frequencies denoted by ω_1 and ω_2, the two-laser-generated optical waves can be mixed at a single photodetector. Assuming the polarization state of the two lasers are the same at the photodetector and that the difference between the two optical frequencies is less than the RF bandwidth of the photodetector, the generated RF photocurrent (I_{RF}) can be expressed as $I_{RF} \propto \gamma[\cos((\omega_1 - \omega_2)t)\cos(\theta_1(t) - \theta_2(t)) - \sin((\omega_1 - \omega_2)t)\sin(\theta_1(t) - \theta_2(t))$, where γ is a proportional factor related to the power of the two lasers and the responsivity of the photodetector $\theta(t)_{1,2}$ and are the time-varying phases of the two optical waves, respectively. From the previous equation, the RF photocurrent will have an oscillating frequency equal to the wavelength spacing between the two lasers. However, the lasers have to be phase-locked to each other. If the phases between the lasers are the same, then the resulting photocurrent is just $I_{RF} \propto \gamma[\cos((\omega_1 - \omega_2)t)]$. If the lasers have a phase difference between them, the mixing process due to photodetection will change the resulting RF photocurrent. Over time, the amplitude will fluctuate, as the phases of the two lasers change from a maximum to a minimum. In order to maximize the RF photocurrent, various phase-locking techniques have been explored. An optical phase lock loop (OPLL) uses an external oscillator to mix with the generated RF signal to create an error signal that is then filtered and used to tune one of the lasers [42]. However, the delay of the loop often has to be less than a nanosecond in order to compensate the phase drifts [43]. Examples of OPLLs that can meet these requirements have been shown [44].

Another method for locking the phases of two lasers is through the optical injection locking process. Much like the injection locking of the OEO discussed previously, the optical injection locking between two lasers allows for the injecting or master laser to impose its phase coherence upon the injected or slave laser. The process is faster than any electronic feedback loop and can create very stable RF sinusoidal signals. An example of an optical injection lock setup is shown in Figure 4.12. The output of the master laser is injected into a slave laser, whose optical isolator has been removed. The master laser imprints its phase variations on the slave laser, locking the slave laser output to the master laser output. By keeping the master laser optical frequency lower than the slave laser, the slave laser will be stably locked to the master laser. The output of the slave laser, containing both the slave and master outputs is then connected to the photodetector. Advanced versions of this type of injection locking have been made, including using multiple slave lasers [45], as well as modulating the master laser to create sidebands that can be used to lock the slave laser [46]. The optical injection-locked system can also be used in an OEO. While most OEO demonstrations use a laser combined with a modulator, the laser can be directly modulated inside

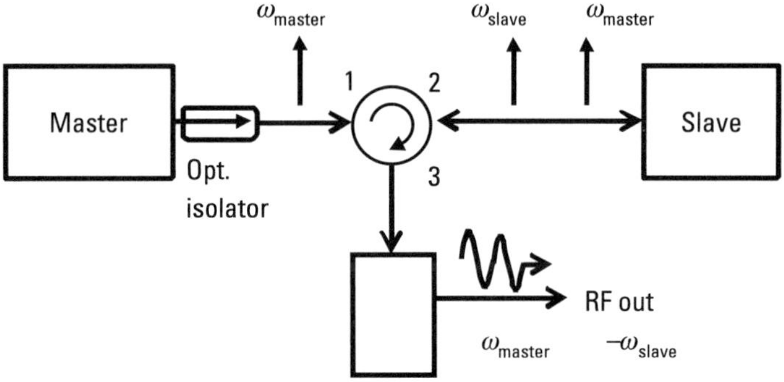

Figure 4.12 Setup for using a master laser to optically injection lock a slave laser in order to generate an RF signal with a frequency equal to the difference between each of the lasers. (After: [47].)

the OEO cavity. In order to increase the oscillating frequency and improve performance, an optically injected OEO (OIL-OEO) has been demonstrated [48]. In this configuration, a slave laser in the OEO cavity is optically injection-locked by a master laser in the stable locking regime. The output of the slave laser is then detected by a photodiode, and the resulting RF signal is then filtered and fed back to the RF input of the slave laser. This allows the OEO to generate a 20-GHz RF signal without the need for any additional modulator. The resulting phase noise is just above –110 dBc/Hz at a 1-kHz offset. Other OEOs have also been demonstrated using optical injection locking of lasers to generate the RF tone [49]. Finally the combination of both the OPLL technique and optical injection locking has been shown to create stable oscillators [50].

One of the advantages to the two-laser system for RF signal generation is the ability to tune the frequency of the signal continuously over a large range and with high speed. While OEOs have been shown that can be tuned in frequency [51, 52], the OEO will only allow frequencies that match one of the cavity modes in the OEO to oscillate. The OEO cavity can be made long, allowing the cavity modes to be spaced very close to one another. However these closely spaced modes will still not allow true continuous tuning. The two-laser system, on the other hand, does not depend on an external cavity and can be continuously tuned over a wide frequency range. The master slave can be tuned either by changing the temperature or drive current of the laser. The temperature will change the length of the optical cavity, tuning the optical frequency of the master laser output. However, temperature tuning is very slow (on the order of seconds) and is not useful for most applications.

Tuning the pump current of the master laser has a much faster response. Yet, changing the drive current will also affect other parameters of the master and/or slave laser. Specifically, the optical power changes. In order to quantify the changes, studies of the dynamics of the optical injection system have been made [53, 54]. One interesting region that has been investigated is the period-one oscillation region. As mentioned previously, when the master laser's optical frequency is lower than the slave laser's optical frequency, stable injection locking is achieved. On the other hand, when the master laser's optical frequency is higher than the slave laser's frequency, a Hopf bifurcation can be crossed, and the period-one oscillation region can be entered. When in the period one regime, the slave laser's relaxation oscillations will become undamped, allowing the slave laser's frequency to shift to lower frequencies [55]. The optical spectrum showing the shift in the slave laser's wavelength is shown in Figure 4.13. This region is very sensitive to the master laser's operating frequency and power. A result of operating in this region is that a very fast tunable oscillator can be realized. Some demonstrations have shown that the RF frequency can be tuned continuously over a range of 5 GHz in approximately 5 ns [56]. This operation appears in Figure 4.14. The use of the period-one oscillation has strong potential for making ultra-fast, continuously widely tunable oscillators.

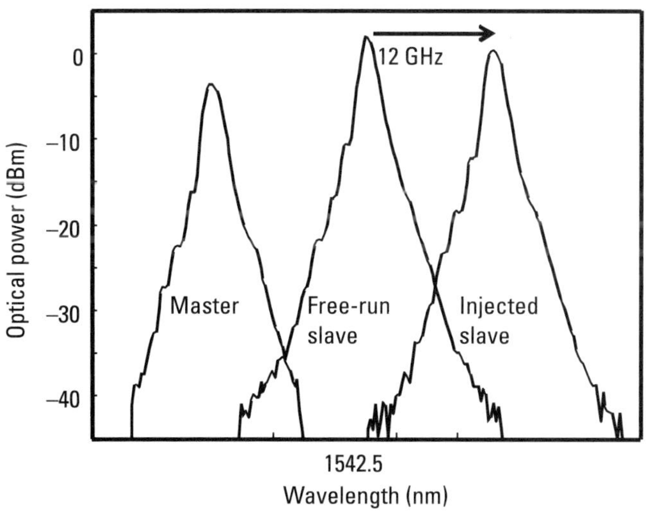

Figure 4.13 Optical spectrum of slave laser when the master laser is off (free-run slave) and when the master is on (injected slave). The period-one locking forces the slave to red-shift.

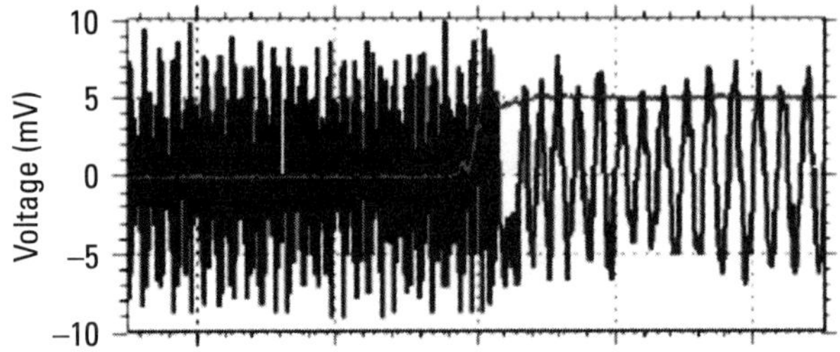

Figure 4.14 Demonstration of switching speed of photonically based oscillator. (After: [47].)

4.5 Conclusions

RF photonic subsystems can be developed to act as oscillators in the demonstration architecture. OEOs can provide very precise clock signals for multiple applications. The demonstrated phase noise performance of an OEO can meet or exceed the performance of other electronic oscillators [28]. The OEO can also be used to synchronize a clock to an incoming RF signal, allowing the signal to be captured on a sampling scope that requires a trigger that matches the frequency of the RF input. Beyond OEOs, the use of two phase-locked lasers can also be used to generate RF clock signals. The two-laser system can be tuned continuously over a large band of frequencies. The use of injection locking of two lasers has also been shown to generate new oscillating tones due to nonlinear dynamics in the laser. Specifically the period-one state has been shown to generate RF signals over a frequency range of 10 GHz, at very high speeds. The ultrafast tunable oscillator based on the period-one dynamics can be used as a local oscillator for downconverting signals or simply to scan through a large spectrum of RF bandwidth. Chapter 5 covers more signal-processing applications using RF photonics, including the oscillators described in this chapter.

References

[1] Vig, J., "Military Applications of High Accuracy Frequency Standards and Clocks," *Transactions on Ultrasonics, Ferroelectrics, and Frequency Control,* Vol. 40, No. 5, 1993, pp. 522–527.

[2] Fukuchi, K., et al. "10.92-Tb/s (273 x 40-Gb/s) Triple-Band/Ultra-Dense WDM Optical-Repeatered Transmission Experiment," *Optical Fiber Communication Conference,* 2001.

[3] Shoop, B., *Photonic Analog-to-Digital Conversion,* New York, NY: Springer Verlag, 2001.

[4] Da Dalt, N., et al., "On the Jitter Requirements of the Sampling Clock for Analog-to-Digital Converters," *Transactions on Circuits and Systems I: Fundamental Theory and Applications,* Vol. 49, No. 9, 2002, pp.1354–1360.

[5] Sartorius, B., "All-Optical Clock Recovery for 3R Optical Regeneration," *Optical Fiber Communication Conference,* 2001.

[6] Clark, T. R., et al., "Phase Noise Measurements of Ultrastable 10 GHz Harmonically Mode Locked Fibre Laser," *Electronics Letters,* Vol. 35, No. 9, 1999, pp. 720–721.

[7] Mule, A. V., et al., "Electrical and Optical Clock Distribution Networks for Gigascale Microprocessors," *Transactions on Very Large Scale Integration (VLSI) Systems,* Vol. 10, No. 5, 2002, pp. 582–594.

[8] Miller, D. A. B. "Rationale and Challenges for Optical Interconnects to Electronic Chips," *Proceedings of the IEEE,* Vol. 88, No. 6, 2000, pp. 728–749.

[9] Harris, B., and B. Trubey, "10G-Enabled Optical Network Architecture Directions for Video, Voice and Data: An MSO Perspective," *Optical Fiber Communication Conference,* 2006.

[10] Redmayne, D., E. Trelewicz, and A. Smith, "Understanding the Effect of Clock Jitter on High Speed ADCs," *Design Note 1013,* Linear Technology Corporation, 2006, http://cds.linear.com/docs/en/design-note/dn1013f.pdf.

[11] Kester, W., *Analog-Digital Conversion,* Analog Devices, 2004. ISBN13: 978-0-91655-027-1.

[12] Devgan, P., "A Review of Optoelectronic Oscillators for High Speed Signal Processing Applications," *International Scholarly Research Notices (ISRN) Electronics,* Vol. 2013, Article ID 401969, 2013.

[13] Navid, R., et al., "Close-In Phase Noise in Electrical Oscillators," *Proc. SPIE Symposium on Fluctuations and Noise*, 2004.

[14] Gheen, K., "Phase Noise Measurement Methods and Techniques," Keysight Webcast, July 2012.

[15] Von der Linde, D. "Characterization of the Noise in Continuously Operating Mode-Locked Lasers," *Applied Physics B,* Vol. 39, No. 4, 1986, pp. 201–217.

[16] Yao, X. S., and L. Maleki, "Optoelectronic oscillator for photonic systems," *Journal of Quantum Electronics,* Vol. 32, No. 7, 1996, pp. 1141-1149.

[17] Yao, X. S., and L. Maleki, "Converting Light into Spectrally Pure Microwave Oscillation," *Optics Letters,* Vol. 21, No. 7, 1996, pp. 483–485.

[18] Yao, X. S., and L. Maleki, "Optoelectronic Microwave Oscillator," *JOSA B,* Vol. 13, No. 8, 1996, pp. 1725–1735.

[19] Neyer, A. and E. Voges, "High-Frequency Electrooptic Oscillator Using an Integrated Interferometer," *Applied Physical Letters,* Vol. 40, No. 1, 1982, pp. 6–8.

[20] Nakazawa, M., T. Nakashima, and M. Tokuda, "An Optoelectronic Self-Oscillatory Circuit with an Optical Fiber Delayed Feedback and Its Injection Locking Technique," *Journal of Lightwave Technology,* Vol. 2. No. 5, 1984, pp. 719–730.

[21] Lewis, M. F., "Novel RF Oscillator Using Optical Components," *Electronics Letters,* Vol. *28,* No. 1, 1992, pp. 31–32.

[22] Lasri, J., et al., "Self-Starting Optoelectronic Oscillator for Generating Ultra-Low-Jitter High-Rate (10 GHz or Higher) Optical Pulses," *Optics Express,* Vol. 11, No. 12, 2003, pp. 1430–1435.

[23] Romisch, S., et al., "Performance Evaluation of an Optoelectronic Oscillator," *Transactions on Ultrasonics, Ferroelectrics, and Frequency Control,* Vol. 47, No. 5, 2000, pp. 1159–1165.

[24] K&L Filter Data Sheet, www.klmicrowave.com.

[25] Yao, X. S., et al., "Dual-Loop Opto-Electronic Oscillator," *Proc. of Frequency Control Symposium,* 1998.

[26] Yao, X. S. and L. Maleki, "Multiloop Optoelectronic Oscillator," *Journal of Quantum Electronics,* Vol. 36, No. 1, 2000, pp. 79–84.

[27] Shumakher, E., and G. Eisenstein, "A Novel Multiloop Optoelectronic Oscillator," *Photonics Technology Letters,* Vol. 20, No. 22, 2008, pp. 1881–1883.

[28] Banky, T., B. Horvath, and T. Berceli, "Optimum Configuration of Multiloop Optoelectronic Oscillators," *JOSA B,* Vol. 23, No. 7, 2006, pp. 1371–1380.

[29] Zhou, W., and G. Blasche, "Injection-Locked Dual Opto-Electronic Oscillator with Ultra-Low Phase Noise and Ultra-Low Spurious Level," *Transactions on Microwave Theory and Techniques,* Vol. 53, No. 3, 2005, pp. 929–932.

[30] Okusaga, O., et al., "Experimental and Simulation Study of Dual Injection-Locked OEOs," *Proc. of Frequency Control Symposium,* 2009.

[31] Okusaga, O., et al., "Spurious Mode Reduction in Dual Injection-Locked Optoelectronic Oscillators," *Optics Express,* Vol. 19, No. 7, 2011, pp. 5839–5854.

[32] Devgan, P., et al., "A Low-Jitter Master-Slave Optoelectronic Oscillator Employing All-Photonic Gain," *Proc. of International Topics on Microwave Photonics,* 2007.

[33] Devgan, P. S., et al., "Improvement in the Phase Noise of a 10 GHz Optoelectronic Oscillator Using All-Photonic Gain," *Journal of Lightwave Technology,* Vol. 27, No. 15, 2009, pp. 3189–3193.

[34] Siegman, A. E., *Lasers,* South Orange, NJ: University Science Books, 1986.

[35] Adler, R., "A Study of Locking Phenomena in Oscillators," *Proc. of the IEEE,* Vol. 61, 1973, pp. 1380–1385.

[36] Lasri, J., and G. Eisenstein, "Phase Dynamics of a Timing Extraction System Based on an Optically Injection-Locked Self-Oscillating Bipolar Heterojunction Phototransistor," *Journal of Lightwave Technology,* Vol. 20, No. 11, 2002, pp. 1924–1932.

[37] Devgan, P., "High Speed Signal Processing Using Nonlinear Fibers and Optoelectronic Devices," Ph.D. Dissertation, Northwestern University, 2006.

[38] Lasri, J., et al., "Ultra-Low Timing Jitter 40Gb/s Clock Recovery Using a Self-Starting Optoelectronic Oscillator," *Photonics Technology Letters*, Vol. 16, No. 1, 2004, pp. 263–265.

[39] Pan, S. and J. Yao, "Optical Clock Recovery Using a Polarization-Modulator-Based Frequency-Doubling Optoelectronic Oscillator," *Journal of Lightwave Technology*, Vol. 27, No. 16, 2009, pp. 3531–3539.

[40] Tsuchida, H., and M. Suzuki, "40-Gbps Optical Clock Recovery Using an Injection-Locked Optoelectronic Oscillator," *Photonics Technology Letters*," Vol. 17, No. 1, 2005, pp. 211–213.

[41] Tsuchida, H., "Subharmonic Optoelectronic Oscillator," *Photonics Technology Letters*, Vol. 20, No. 17, 2008, pp. 1509–1511.

[42] Langley, L., et al., "Packaged Semiconductor Laser Optical Phase-Locked Loop (OPLL) for Photonic Generation, Processing and Transmission Of Microwave Signals," *Transactions on Microwave Theory and Techniques*, Vol. 47, No. 7, 1999, pp. 1257–1264.

[43] Seeds, A., and K. Williams, "Microwave Photonics," *Journal of Lightwave Technology*, Vol. 24, No. 12, 2006, pp. 4628–4641.

[44] Gliese, U., et al., "A Wideband Heterodyne Optical Phase-Locked Loop for the Generation of 3–18 GHz Microwave Carriers," *Photonics Technology Letters*, Vol. 4, No. 8, 1992, pp. 936–938.

[45] Goldberg, L., et al., "Microwave Signal Generation with Injection Locked Laser Diodes," *Electronics Letters*, Vol. 19, No. 13, 1983, pp. 491–493.

[46] Schneider, G., et al., "Radiofrequency Signal-Generation System with Over Seven Octaves of Continuous Tuning," *Nature Photonics*, Vol. 7, No. 2, 2013, pp. 118–122.

[47] Devgan, P., "Signal Processing Utilizing Radio Frequency Photonics," *AFRL Report*, AFRL-RY-WP-TR-2017-0172, 2017

[48] Sung, H.-K., et al., "Optoelectronic Oscillators Using Direct-Modulated Semiconductor Lasers Under Strong Optical Injection," *Journal of Selected Topics in Quantum Electronics*, Vol. 15, No. 3, 2009, pp. 572–577.

[49] Haji, M., et al., "High Frequency Optoelectronic Oscillators Based on the Optical Feedback of Semiconductor Mode-Locked Laser Diodes," *Optics Express*, Vol. 20, No. 3, 2012, pp. 3268–3274.

[50] Johansson, L., and A. Seeds, "Millimeter-Wave Modulated Optical Signal Generation with High Spectral Purity and Wide Locking Bandwidth Using a Fiber-Integrated Optical Phase-Lock Loop," *Photonics Technology Letters*, Vol. 12, No. 6, 2000, pp. 690–693.

[51] Tang, Z., et al., "Tunable Optoelectronic Oscillator Based on a Polarization Modulator and a Chirped FBG," *Photonics Technology Letters*, Vol. 24, No. 17, 2012, pp. 1487–1489.

[52] Li, W., and J. Yao, "An Optically Tunable Optoelectronic Oscillator," *Journal of Lightwave Technology,* Vol. 28, No. 18, 2010, pp. 2640–2645.

[53] Simpson, T., "Mapping the Nonlinear Dynamics of a Distributed Feedback Semiconductor Laser Subject to External Optical Injection," *Optics Communications,* Vol. 215, No. 1, 2002, 2002, pp. 135–151.

[54] Wieczorek, S., et al., "The Dynamical Complexity Of Optically Injected Semiconductor Lasers," *Physics Reports,* Vol. 416, No. 1, 2005, pp. 1–128.

[55] Hwang, S., J. Liu, and J. White, "Characteristics of Period-One Oscillations in Semiconductor Lasers Subject to Optical Injection," *Journal of Selected Topics in Quantum Electronics,* Vol. 10, No. 5, 2004, pp. 974–981.

[56] Usechak, N., J. Suelzer, and J. Haefner, "High-Speed Wideband Voltage-Controlled Oscillator via an Injection-Locked Laser," *Photonics Technology Letters,* Vol. 29, No. 13, 2017, pp. 1132–1135.

5

Signal Isolation Utilizing RF Photonics

In Chapters 1–3, the demonstration architecture has only considered processing incoming RF signals from the antenna. However, the system can also be used to transmit RF signals. One possible architecture could include two antennas, one that is used to transmit signal and another used to receive. A more elegant solution, however, would use a single antenna for both transmission and reception. To accomplish both transmit and receive functions with a single antenna, the transmission path must be isolated from the receive path. This is accomplished with an RF circulator, which is connected right after the antenna, as shown in Figure 5.1. The three-port circulator device isolates the transmit path to the antenna from the receive path. On the receive side, the architecture will often need to be able to sense only a certain part of the overall spectrum. Being able to filter out specific signals or sets of signals can be useful. The receive path has a new subsystem, which can separate multiple received signals before processing. RF photonics can provide solutions for both the RF circulator and filters. The photonic-based filters can be widely tunable and have narrow frequency bandwidths. The photonic-based circulator can provide high isolation between transmit and receive paths. This chapter covers isolation and separation of RF signals, providing examples of how RF photonic applications can address these needs.

5.1 Need for Signal Separation

The RF spectrum is increasingly being filled with new signals. While traditional amateur transoceanic radio communications operated in the HF band

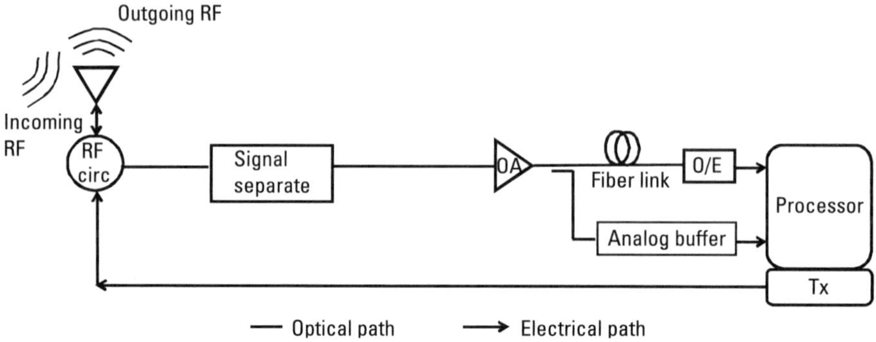

Figure 5.1 Demonstration architecture with RF circulator and RF signal separation subsystems.

(known as HAM radio) and analog broadcast television used the very high frequency (VHF) and lower ultrahigh frequency (UHF) bands, other signals began to use these frequency bands as well. Frequencies in the VHF band (118–136 MHz) are used for air traffic control as well as for civil emergency radio communications (138–144 MHz) [1]. Extending into the UHF band, the 400-MHz frequency is used for time and frequency standard transmission to satellites. Cordless landline telephones use the 900-MHz frequency band but have extended into the 2.5- and 5-GHz bands. Equivalently, the use of home Wi-Fi routers has also made use of these same frequency bands for connecting computers, tablets, and smart phones to the internet. The 1.2- and 1.5–1.6-GHz frequencies are used for commercial GPS signals. The 1.675-GHz frequency is used for meteorological satellite data, while the 2.7–2.9 GHz band is used for airport surveillance radars. Looking beyond the UHF band, the 4.2–4.4-GHz frequency band is used for airborne radar altimeters. The 9–9.5-GHz range is used for precision approach radars as well as airborne weather radars. Speed guns operate at 10.525 GHz, as well as 24.25 GHz. The 15.7–16.2-GHz band is used for airport surface detection equipment, and the automotive industry uses the 76–77-GHz band for collision and road radars [1]. In general, the NTIC long-range spectrum [2] use plan predicts that the 2–18-GHz frequency spectrum will continue to be used with increases in use from 8.5 to 10 GHz and 14 to 15.5 GHz. At higher frequencies, the increase will be seen in the 31–36-GHz band for 5G communication as well as the 77-GHz range for automotive radar and road sensing. As the congestion of signals in these frequency bands increases, better isolation will be needed. To accomplish this, the first consideration would be the use of RF filters, specifically widely tunable filters that also have narrower passbands than those of traditional electronic filters.

5.2 Using RF Photonics for Separation of Signals

As seen in Figure 5.1, the receive path of the architecture has a new subsystem for signal separation. The subsystem permits certain signals to continue to the processing block while removing others. In essence the subsystem acts as an RF filter. The easiest example of an RF filter using photonics is simply the use of an optical filter. When using an optical filter, the modulation format has to be considered. Since most modulation formats are dual-sideband, the bandwidth of the optical filter will affect the RF bandwidth. An example is shown in Figure 5.2. Assuming an optical filter with a given bandwidth, the dual-sideband modulation will only pass through RF signals of one-half the filter bandwidth. As seen in Figure 5.2(a), the optical carrier ideally sits in the middle of the optical filter, so that the sidebands see the same filter response. The use of dual-sideband modulation can be a benefit as it allows an optical filter of twice the needed bandwidth to be used. On the other hand, if the full bandwidth of the optical filter is necessary, single-sideband modulation can be used, as seen in Figure 5.2(b). In either case, the modulation format will play a role in the bandwidth of the optical filter that can be used. Note that most optical filters are based on a cavity and thus have a periodic structure in their response. The passband will repeat at a frequency spacing known as the free spectral range (FSR) [3]. Often, the narrower the passband, the narrower the FSR. The FSR can actually limit the wavelength range over which the filter will be usable and has to be taken into consideration.

The most common metric for the filter is the quality factor, also known as the Q-factor. The Q factor is defined as $Q = F_c/\Delta F$, where F_c is the center

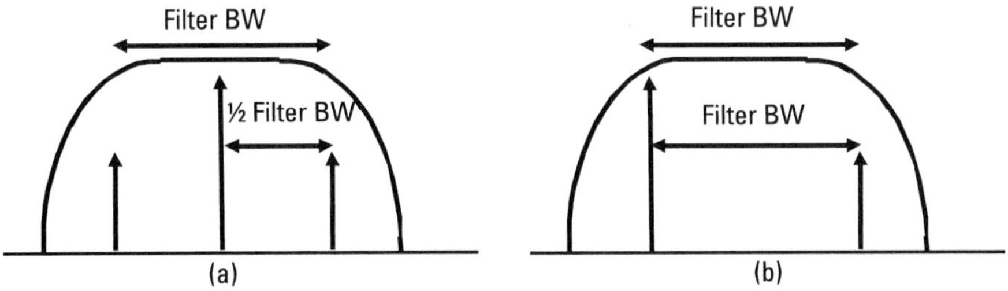

Figure 5.2 Optical filter with either (a) dual-sideband modulation (left) or (b) single-sideband modulation (right). (After: [4].)

frequency of the filter, and ΔF is the full width at half-maximum bandwidth of the filter. For typical electronic-based filters, Q factors between 1,000 and 10,000 are easily achievable. At a center frequency of 10 GHz, a bandwidth between 100 and 1,000 MHz can be realized. Often, these can be too wide for the chosen operation, and narrower bandwidth filters are required. Optical filters can provide much higher Q factors than electronic solutions.

Optical filters have been realized in many different ways, including using SBS [5], ring resonators [6], and thin-film materials [7]. Nevertheless, here we focus on the fiber Bragg grating filter. The Bragg grating filter is realized in optical fibers by writing a grating onto a section of optical fiber. The grating is usually written on the optical fiber. Changes in the refractive index of the fiber create a periodic grating structure in the fiber. Light of a certain wavelength will constructively add up as a reflection off of the grating. Other wavelengths will not constructively reflect and will pass through the grating as seen in Figure 5.3. This allows the filter to act as either a bandpass (in reflection) or a notch filter (in transmission). The wavelength that reflects meets the condition $\Lambda = \lambda/2$ where λ is the wavelength of the reflected light and Λ is the period of the grating. Fiber Bragg gratings are commercially available, with bandwidths ranging from 50 MHz to 50 GHz. One of the drawbacks to these types of filters are their inability to easily tune the center frequency. In order to change the center frequency, the spacing between the grating lines of the Bragg filter has to be changed. Changing the spacing can be accomplished by either stressing the fiber or changing the temperature of the fiber [8]. In either case, the idea is to physically stretch or compress the fiber so the spacing of the grating lines change. This in turn changes the wavelength that will reflect at the grating as given by the relationship above. Stretching the fiber is a slow process and can only be done over a limited amount before causing damage to the fiber. In order to overcome this limitation, other filter techniques have been investigated.

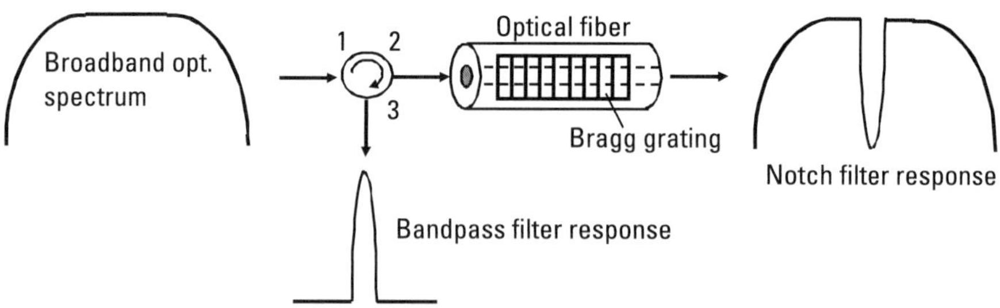

Figure 5.3 A fiber Bragg grating with a broadband optical source at the input. The transmitted response acts like a notch filter while the reflected response is a bandpass filter. (After: [4].)

5.3 Finite Impulse Response Filters Using RF Photonics

A more advanced method for generating an RF filter using RF photonics is the use of delay lines. Unlike the delay lines used for analog storage as described in Chapter 4, the delay lines are used to create time delays between the same RF signal. This creates the equivalent of a finite impulse response (FIR) filter [9]. Before discussing the optical implementation, a brief discussion of the FIR filter is warranted. For linear, time-invariant systems, the output of the system $y(t)$ can be calculated by taking a continuous time convolution $y(t)=\int_{-\infty}^{\infty} h(\tau)s(t-\tau)\ d\tau$ evaluated from time $t=-\infty$ to ∞ between the input signal $s(t)$ and the impulse response of the system $h(t)$. The convolution operation can also be done in the discrete time domain. Discrete time signals are time-sampled versions of continuous time signals. Figure 5.4(a) shows a continuous time varying signal, and Figure 5.4(b) shows the discrete time version, where T is the sample time period. Discrete time signal processing is used in computers to perform operations, such as filtering signals. The equivalent convolution in the digital domain is the discrete convolution sum $y[n]=\sum_{m=-\infty}^{\infty} h[m]s[n-m]$ [10]. In this case the integral is replaced with a summation of the input signal $s(t)$ sampled at T, yielding $s[n]$, where n is the total number of samples. The discrete impulse response is denoted as $h[n]$, which is also a sampled version of $h(t)$ at a sample rate of T. When the number of samples is infinite, the system is said to have an infinite impulse response (IIR). Since practical implementations cannot be infinite, a finite number of samples is used. This is now called a FIR system. As suggested earlier, the simplest example of signal processing is providing a filtered response

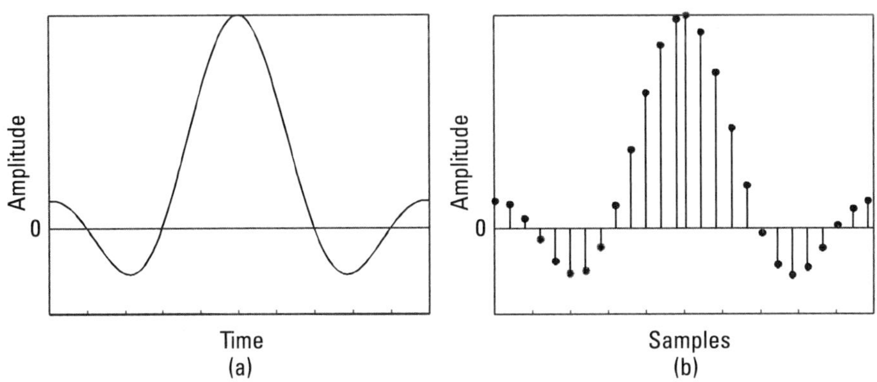

Figure 5.4 (a) A continuous time plot of the sinc function and (b) a sampled version of the same sinc function in (a). (After: [4].)

of an input signal. One of the easiest examples of filtering is a low-pass filter, in which only signals lower than a cutoff frequency pass through without modification, while signals of a higher frequency are forced to zero. In order to realize a practical implementation of this type of filter, a FIR filter can be used. The mathematical expression is given here $y(t) = \sum_{m=0}^{N-1} h[m]s[t - mT]$ [9]. The FIR filter is simply the discrete convolution sum of the sampled impulse response of a given filter shape with multiple time-delayed versions of the signal. An example is shown in Figure 5.5. A sinc function in the time domain will have a low-pass filter response in the frequency domain. The sinc function is defined as $\text{sinc}(t) = \sin(\pi t)/(\pi t)$. The sampled version of the sinc function is shown in Figure 5.5(a). The sampled points of the sinc are used as filter coefficients $h[m]$ in the FIR convolution sum. The number of samples is often referred to as the number of taps in the FIR filter. When the input is an impulse, the resulting frequency response is plotted in Figure 5.5(b), with N = 40. The shape of the filter is that of a low-pass filter. Yet, it does not match the ideal response. Clearly, this is because only a finite sample of the sinc function was used. Regardless, this is a way to realize a filter response with a finite number of elements or taps.

In order to realize an FIR filter in practice, three operations need to be performed. First, the incoming signal needs to be replicated N-number of times. Each of the replicas then needs to be time-delayed and attenuated. Finally, the N-number of time-delayed signals has to be summed together. This accomplishes in practice the convolution sum of the FIR filter. Photonic implementations of an FIR filter have the advantage of providing time delays that can be very low-loss as discussed previously. Photonics also provides the use of multiple wavelengths that can be modulated by the same RF signals

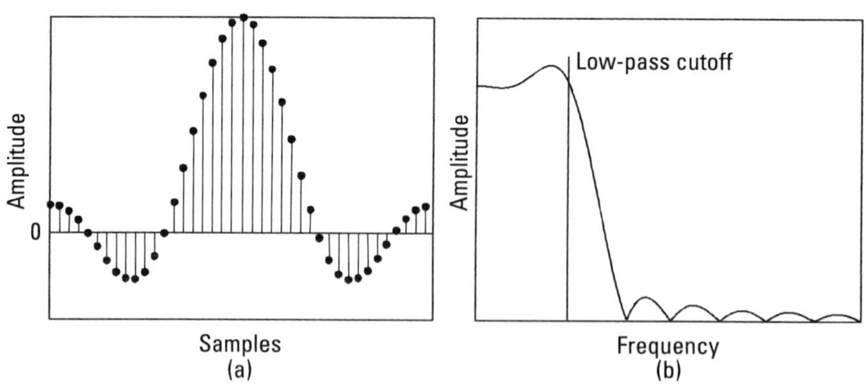

Figure 5.5 (a) A sampled sinc function with 40 samples and (b) the resulting low-pass filter response with an impulse input using the sinc coefficients from (a).

and then split to multiple paths. A photonic-based FIR filter can implement this in a couple of different ways.

Figure 5.6 illustrates the first example. The output of an optical source, consisting of multiple, distinct wavelength, optical carriers, is connected to an optical modulator to imprint the RF waveform onto each of the wavelengths. The output of the optical modulator is then connected to a wavelength demultiplexer. The demultiplexer separates the different wavelengths into parallel paths. Each of the paths is then attenuated and passed through a delay line. Each of the delay lines is one time period, *T*, longer than the one before it. At the end, the different paths are then recombined using a wavelength multiplexer, whose output is connected to a photodiode. The photodiode then takes the different wavelengths and sums up the RF signals. This accomplishes the convolution sum of the FIR filter. The disadvantage of this architecture is that each of the fiber delay lengths has to be maintained relative to the other delays. External environmental perturbations could cause the lengths of the different fibers to change independent of each other, essentially changing the sample time of each tap of the filter. A phase shifter can be added after each attenuator to correct environmental perturbations, as seen in Figure 5.6. On the other hand, an alternative architecture can address this issue. In Figure 5.7, the same multiwavelength optical source and modulator is used. Instead of separating the wavelengths first, they all pass through the same optical fiber. The fiber has multiple Bragg gratings written into it, like the one shown in the Section 5.2. Each of the Bragg gratings is set to reflect a specific wavelength of the optical source. Each of the Bragg gratings is also spaced by a length that yields a delay of $T/2$. In this way, the first wavelength sees a delay of *T*, the second wavelength sees a delay of $2T$, the third wavelength sees a delay of $3T$, and so on. All of the reflected wavelengths are then collected onto a single photodiode for the addition operation. To weight each of the signals with the correct impulse response coefficient, the reflectivity of the Bragg grating can be changed [11], or each wavelength's optical power

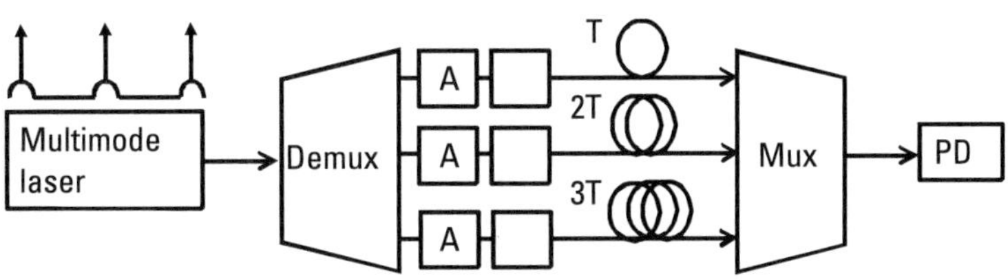

Figure 5.6 An RF photonic implementation of an FIR filter using multiwavelength source and different time delays. (After: [4].)

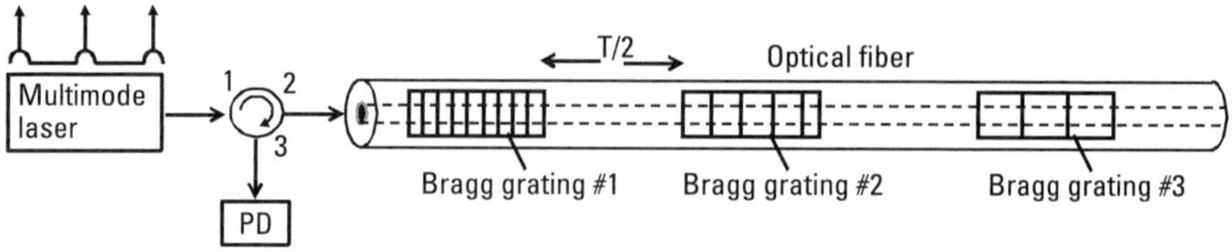

Figure 5.7 An RF photonic implementation of an FIR filter using a multiwavelength source and different Bragg gratings. (After: [4].)

can be set at the optical modulator. The single-path implementation of the FIR filter provides a single path for all the wavelengths, overcoming the parallel path architecture's sensitivity to external influences. Still, this advantage comes at the expense of not being able to change the weights of each time delay as easily as in the parallel path architecture.

There are a few caveats that must be discussed when discussing the photonic implementation of an FIR filter. Ideally, the FIR filter acts to add up the different time-delayed and amplitude-weighted versions of the RF waveform. However, the photonic implementation of the FIR filter still has the optical carrier. The optical carrier will also see the time delays and weights and will interfere at the photodetector, degrading the performance of the FIR filter in the RF domain. Thus, the use of either a multiwavelength source, as seen in Figures 5.6 and 5.7, or a broadband optical source is necessary to achieve the desired FIR filter response [9]. Another consideration is the weighting function that is to be implemented for the FIR filter. Looking at the sampled sinc function in Figure 5.5(a), the samples consist of both positive and negative numbers. In practice, photonic implementations of the FIR filters cannot provide negative weights to the time delays. Work has been shown to allow for negative weights [12], including using phase shifts in the modulator to implement negative weights [13], using differential photodetection [14] and other methods [15–17]. Unfortunately, these require more complex setups than the simple implementations shown here. One last consideration is that the FIR filter is a periodic filter. The response of the filter will repeat with a separation denoted by the FSR. This is a consequence of the sampling nature of the FIR filter implementation. The FSR is related to the sample time period T by the simple relation $FSR = 2\pi / T$. Multiple implementations have been investigated to generate filter responses that do not repeat (i.e., single-passband filters), including cascading filters of different FSRs that have a single passband in common [18], or a combination of an asymmetric Mach-Zehnder interferometer (AMZI) and an optical modulator to select only one mode of the AMZI [19].

The FIR filters have been shown to be very flexible and adaptable. The weights at each of the time delays can be set by variable optical attenuators. Therefore, the weights can be dynamically changed. This will allow an FIR filter to go from a low-pass response to a high-pass response simply by flipping every other coefficient in the filter weights (also known as spectral inversion) [20]. In addition, the center frequency of the FIR filter response can be tuned. Tuning is often achieved by the use of a tunable laser source combined with a chirped fiber Bragg grating [21, 22]. In either case the change in wavelength will result in different time delays for each of the delay lines, resulting in different center frequencies for the filter. Since the wavelength of the laser can be tuned very quickly, the FIR filter can be tuned at much faster speeds than the fiber Bragg grating described earlier. Many other methods have also been investigated for making tunable frequency FIR filters [23–25].

5.4 Isolation of RF Signals Along a Common Path

Clearly, there are many examples of separating RF signals using photonic methods. Different filter architectures can be used to cut out certain signals from a larger spectrum, allowing them to be processed without being affected by neighboring signals. Beyond separation of signals, another operation that is of use is the isolation of transmit and receive signals at a single antenna. Multiple applications require the capability to both transmit and receive from the same antenna. Examples include radar, radio communication, and wireless routers. One method commonly used is a time-gating method. Time gating means that the antenna is set for a certain time in transmission mode and then switched to reception mode. High-speed T/R, or transmit/receive, switches have been used in time-gating architectures [26]. While this works for some applications, the time-gating method will have dead times, where signals of interest are not being received. For certain applications, the dead time is unacceptable.

To overcome this limitation, an RF circulator can be used to isolate the transmitted and received signals from the same antenna. Figure 5.8 shows a close-up of the RF circulator pictured in Figure 5.1. As mentioned previously, the RF circulator is typically a three-port device, where signals to be transmitted enter port 1 and are passed onto port 2 with very low loss (~1 dB). Port 2 is connected to the antenna, where signals can both be transmitted and received simultaneously, without any time gating. The signals that are received enter port 2 and are passed to port 3 with low loss (~1–2 dB). The losses between ports 1 and 3 are very high (> 20 dB), as well as the loss for any signal entering port 2 and leaking out to port 1 (> 25 dB). RF circulators provide

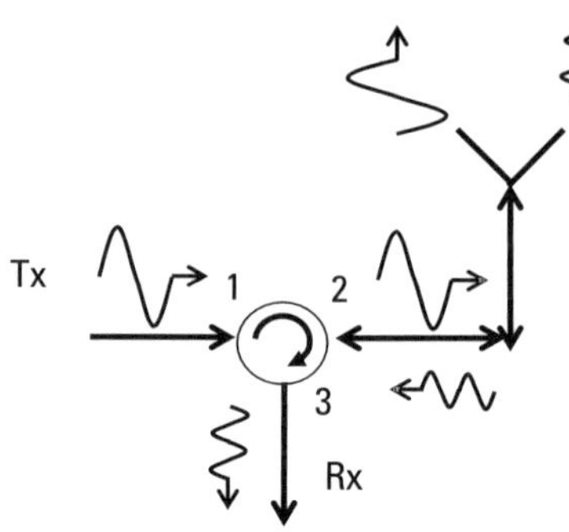

Figure 5.8 An RF circulator connected to an antenna to transmit and receive simultaneously.

very good isolation between the receive and transmit paths. Unfortunately, ferrite-based RF circulator technology cannot easily cover wide frequency ranges while maintaining high isolation. Wideband RF circulators have been demonstrated that provide an isolation of 10 dB at frequencies above 5 GHz to just under 20 GHz [27]. At higher frequencies, the isolation begins to degrade while the insertion loss increases. Methods have been investigated to improve the isolation by using either multiple circulators [28] or matched pairs of circulators with antennas [29].

In order to provide larger isolation of the transmit and receive paths from a single antenna over a wide frequency range, RF photonics can provide a solution. Combining a bidirectional signal interface using an optical modulator [30] with single-sideband modulation multiplexing [31] allows for high isolation between two counterpropagating RF signals traveling on a common path [32]. As shown in Figure 5.9, the configuration acts like a three-port RF circulator. The RF signal for transmission is input to port 1.

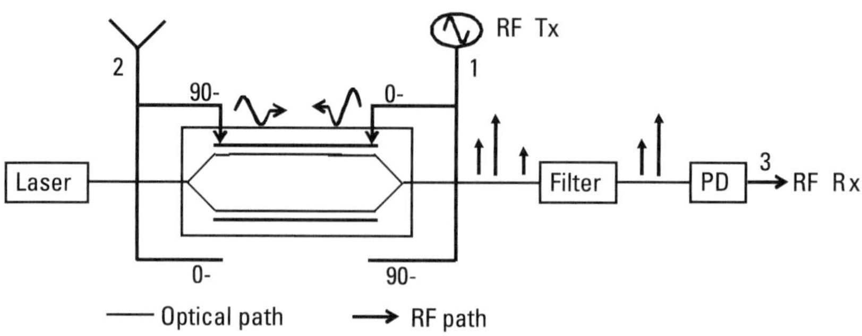

Figure 5.9 An RF photonic-based circulator for isolating transmit and receive paths. (After: [32].)

The transmit signal is then split by a 90° hybrid onto the two RF waveguide electrodes in the MZM, with the zero-degree output on the top waveguide electrode and the 90° output on the bottom waveguide electrode. Note that the RF signal from port 1 counterpropagates with respect to the optical field, which is input to the MZM from the laser at the optical input of the MZM. The split signals then propagate to another 90° hybrid, which is placed in a complementary arrangement to the first hybrid. For this case, the 90° input of the second hybrid is connected to the top RF electrode waveguide, and the zero-degree input is connected to the bottom RF electrode waveguide. The recombined RF signal is then transmitted out port 2 through the antenna. The antenna can also receive signals at the same time. Thus, on port 2, another RF signal can be received. The received signal is split by the second 90° hybrid, with the 90° output on the top RF electrode waveguide and the zero-degree output on the bottom RF electrode waveguide. Note that the received RF signal will copropagate with the optical field of the laser in the MZM and counterpropagate to the transmitted RF signal. The recombined received signal can be transmitted out port 1. Both RF signals will be upconverted onto the optical carrier frequency of the laser. With the correct DC bias chosen for each electrical waveguide, the resulting optical output will be single sideband-modulated. Since the transmit RF signal passes through the complementary 90° hybrid as compared to the received RF signal, the transmit RF signal will appear on the opposite side of the optical carrier as compared to the received RF signal. The optical output of the MZM is then passed through an optical filter in order to remove the unwanted transmit RF signal's sideband, while preserving the sideband with the received RF signal. The output of the filter is connected to a photodetector where the recovered RF signal is then output at port 3.

A mathematical description of the single-sideband modulation is shown in (5.1). The MZM has a transfer function given by

$$E_{out}(t) = \frac{1}{2}\left[e^{i\varphi_1(t)} - e^{i\varphi_2(t)}\right]E_{in}(t) \tag{5.1}$$

In the case of the RF signal entering port 2, the phase modulation on each arm of the MZM after the 90° hybrid can be written as $\varphi_1(t) = \varphi_{dc1} + \varphi_{rf1}\sin(\Omega_{rf}t)$ and $\varphi_2(t) = \varphi_{dc2} + \varphi_{rf2}\cos(\Omega_{rf}t)$, with $\varphi_{dc1,2} = \pi(V_{dc1,2}/V_{\pi,dc1,2})$ and $\varphi_{rf1,2} = \pi(V_{rf1,2}/V_{\pi,rf1,2}(\Omega_{rf}))$.Now using the Jacobi Anger expansion $e^{iz\cos\theta} = \sum_{n=-\infty}^{\infty} i^n J_n(z)e^{in\theta}$ and $e^{iz\sin\theta} = \sum_{n=-\infty}^{\infty} J_n(z)e^{in\theta}$ the following can be written

$$E_{carrier}(t)=\frac{\overline{E}_{in}e^{i\omega_o t}}{2}\left[e^{i\varphi_{dc1}}J_0(\varphi_{rf1})-e^{i\varphi_{dc2}}J_0(\varphi_{rf2})\right],$$
$$E_{usb,fund}(t)=\frac{\overline{E}_{in}e^{i\omega_o t-i\Omega_{rf}t}}{2}\left[-e^{i\varphi_{dc1}}J_1(\varphi_{rf1})-ie^{i\varphi_{dc2}}J_1(\varphi_{rf2})\right], \quad (5.2)$$
$$E_{lsb,fund}(t)=\frac{\overline{E}_{in}e^{i\omega_o t+i\Omega_{rf}t}}{2}\left[e^{i\varphi_{dc1}}J_1(\varphi_{rf1})-ie^{i\varphi_{dc2}}J_1(\varphi_{rf2})\right],$$

In this case, the RF signal is the same on both electrical waveguides so $\varphi_{rf1} = \varphi_{rf2}$. For the condition of single-sideband operation, there are two cases. For the upper sideband to be nulled, $\varphi_{dc1} = -\pi/2 + \varphi_{dc2}$. For the lower sideband to be nulled, $\varphi_{dc1} = \pi/2 + \varphi_{dc2}$. Now for the RF signal input to port 2, the hybrid is in the complementary configuration so $\varphi_1(t) = \varphi_{dc1} + \varphi_{rf1}\cos(\Omega_{rf}t)$ and $\varphi_2(t) = \varphi_{dc2} + \varphi_{rf2}\sin(\Omega_{rf}t)$. In this configuration, the resulting fields are written as

$$E_{carrier}(t)=\frac{\overline{E}_{in}e^{i\omega_o t}}{2}\left[e^{i\varphi_{dc1}}J_0(\varphi_{rf1})-e^{i\varphi_{dc2}}J_0(\varphi_{rf2})\right],$$
$$E_{usb,fund}(t)=\frac{\overline{E}_{in}e^{i\omega_o t-i\Omega_{rf}t}}{2}\left[ie^{i\varphi_{dc1}}J_1(\varphi_{rf1})+e^{i\varphi_{dc2}}J_1(\varphi_{rf2})\right], \quad (5.3)$$
$$E_{lsb,fund}(t)=\frac{\overline{E}_{in}e^{i\omega_o t+i\Omega_{rf}t}}{2}\left[ie^{i\varphi_{dc1}}J_1(\varphi_{rf1})-e^{i\varphi_{dc2}}J_1(\varphi_{rf2})\right],$$

Now for the single-sideband condition, the upper sideband is nulled when $\varphi_{dc1} = \pi/2 + \varphi_{dc2}$, and the lower sideband is nulled when $\varphi_{dc1} = -\pi/2 + \varphi_{dc2}$. From this one can see that the RF signals will always appear on opposite sides of the optical carrier when the bias conditions are set for single-sideband modulation. This allows for the RF signals to be separated in the optical spectrum from each other. Then, using an optical filter will allow for one sideband to be removed while the carrier and the other sideband will pass on to the photodetector to recover the RF signal at port 3.

The RF circulator using photonics has many advantages compared to the ferrite-based circulator, beyond just the increased RF bandwidth. Because the transmit signal is removed by the optical filter before the detector, the RF isolation between port 1 and port 3 can be infinite in theory. The insertion loss between port 1 and port 2 can be relatively low, as the signals are split and then recombined between the two ports. The MZM has been shown to handle RF powers as high as 20W when used in a bidirectional application [30]. However, there are a few drawbacks to this solution. Since most of the ferrite-based RF circulators are passive devices, they only add loss between ports. However, the photonic-based RF circulator is an active device and thus adds noise and distortion. The system from port 2 to port 3 is essentially an RF photonic link. Therefore it suffers from the drawbacks of a traditional RF photonic link. The loss, NF, and linearity of the system from port 2 to port 3 will be determined by the laser RIN and optical power, as well as the modulation efficiency (V_π) of the MZM, as shown in Chapter 2. The application will decide what performance is needed for the receive link.

5.5 Conclusions

RF signal separation and isolation can play a significant role in different applications. As the RF spectrum continues to be exploited for various uses, the number of signals will increase over time. The ability to separate signals of interest from unwanted signals will continue to be a necessary function for most architectures. Beyond simply separating signals, the FIR filters presented in this chapter can be used to recover specific types of signals. By knowing the specific encoding of certain signals, the FIR system can be used as a matched filter to recover and decode these signals. In a congested RF environment, these types of encoded signals will be useful to maximize spectral efficiency. The cellular phone networks already use these types of codes to enable multiple users to access the service in a limited frequency bandwidth [33]. In addition, being able to both transmit and receive signals from a single antenna is also of great benefit. Often, physical space or cost limits a system to only a single antenna. While T/R switches can be used to time-gate the transmit and receive signals, the ultimate solution would allow for simultaneous use of the antenna for both transmission and reception. The RF circulator provides the capability to both transmit and receive RF signals by providing isolation between the transmit and receive paths. While the traditional ferrite-based RF circulators can provide isolation of signals around 10 dB, RF photonic solutions have shown isolation between 20 and 40 dB over equivalent operational frequencies. Clearly, the RF photonic circulator can overcome the limitations of traditional circulator technologies.

References

[1] Stine, J., and D. Portigal, "Spectrum 101: An Introduction to Spectrum Management," MTR 04W0000048, MITRE, 2004.

[2] U.S. Spectrum Requirements, Projections and Trends, Ch. 4, www.ntia.doc.gov/legacy/osmhome/EPS/openness/sp_rqmnts/radar4/html.

[3] Hecht, E., *Optics,* Boston, MA: Addison-Wesley, 1998.

[4] Devgan, P., "Signal Identification and Isolation Utilizing Radio Frequency Photonics," *AFRL Report,* AFRL-RY-WP-TR-2017-0158, 2017.

[5] Tanemura, T., Y. Takushima, and K. Kikuchi, "Narrowband Optical Filter, with a Variable Transmission Spectrum, Using Stimulated Brillouin Scattering in Optical Fiber," *Optics Letters,* Vol. 27, No. 17, 2002, pp. 1552–1554.

[6] Barwicz, T., et al., "Microring-Resonator-Based Add-Drop Filters in SiN: Fabrication and Analysis," *Optics Express,* Vol. 12 No. 7, 2004, pp. 1437–1442.

[7] MacLeod, H. A., *Thin-Film Optical Filters,* Boca Raton, FL: CRC Press, 2001.

[8] Hill, K. O., and G. Meltz, "Fiber Bragg Grating Technology Fundamentals and Overview," *Journal of Lightwave Technology,* Vol. 15, No. 8, 1997, pp. 1263–1276.

[9] Capmany, J., B. Ortega, and D. Pastor, "A Tutorial on Microwave Photonic Filters," *Journal of Lightwave Technology,* Vol. 24, No. 1, 2006, pp. 201–229.

[10] Oppenheim, A. V., and W. Schafer, *Discrete Time Signal Processing,* Upper Saddle River, NJ: Prentice Hall, 1989.

[11] Yao, J., "A Tutorial on Microwave Photonics," *IEEE Photonics Society Newsletter,* 2012.

[12] Minasian, R., E. Chan, and X. Yi, "Microwave Photonic Signal Processing," *Optics Express,* Vol. 21, No. 19, 2013, pp. 22918–22936.

[13] Capmany, J., et al., "Microwave Photonic Filters with Negative Coefficients Based on Phase Inversion in an Electro-optic Modulator," *Optics Letters,* Vol. 28, No. 16, 2003, pp. 1415–1417.

[14] Sales, S., et al., "Experimental Demonstration of Fibre-Optic Delay Line Filters with Negative Coefficients," *Electronics Letters,* Vol. 31, No. 13, 1995, pp. 1095–1096.

[15] Coppinger, F., et al., "All-Optical RF Filter Using Amplitude Inversion in a Semiconductor Optical Amplifier," *Transactions on Microwave Theory and Techniques,* Vol. 45, No. 8, 1997, pp. 1473–1477.

[16] You, N., and R. A. Minasian, "Synthesis of WDM Grating-Based Optical Microwave Filter with Arbitrary Impulse Response," in *Proc. of International Topical Meeting on Microwave Photonics,* 1999.

[17] Loayssa, A., et al., "Demonstration of Incoherent Microwave Photonic Filters with All-Optical Complex Coefficients," *Photonics Technology Letters,* Vol. 18, No. 16, 2006, pp. 1744–1746.

[18] You N., and R. Minasian, "High-Q Optical Microwave Filter," *Electronics Letters,* Vol. 35, No. 24, 1999, pp. 2125–2126.

[19] Capmany, J., et al., "High-Q Microwave Photonic Filter with a Tuned Modulator," *Optics Letters,* Vol. 30, No. 17, 2005, pp. 2299–2301.

[20] Smith, S., *The Scientists and Engineers Guide to Digital Signal Processing,* 1997, www.DSPguide.com.

[21] Zhang, W., et al., "Fibre Optic Radio Frequency Notch Filter with Linear And Continuous Tuning by Using a Chirped Fibre Grating," *Electronics Letters,* Vol. 34, No. 18, 1998, pp. 1770–1772.

[22] Hunter, D. A., and R. Minasian, "Tunable Transversal Filter Based on Chirped Gratings," *Electronics Letters,* Vol. 31, No. 25, 1995, pp. 2205–2207.

[23] Yan, Y., and J. Yao, "A Tunable Photonic Microwave Filter with a Complex Coefficient Using an Optical RF Phase Shifter," *Photonics Technology Letters,* Vol. 19, No. 19, 2007, pp. 1472–1474.

[24] Frankel, M., and R. Esman, "Fiber-Optic Tunable Microwave Transversal Filter," *Photonics Technology Letters,* Vol. 7, No. 2, 1995, pp. 191–193.

[25] Mora, J., et al., "Photonic Microwave Tunable Single-Bandpass Filter Based on a Mach-Zehnder Interferometer," *Journal of Lightwave Technology,* Vol. 24, No. 7, 2006, pp. 2500–2509.

[26] Skolnik, M., *Radar Handbook,* New York, NY: McGraw-Hill Education, 2008.

[27] Schloemann, E. F., "Circulators for Microwave And Millimeter-Wave Integrated Circuits," *Proceedings of the IEEE,* Vol. 76, No. 2, 1988, pp. 188–200.

[28] Ho, T. Q., et al., "Wideband High Isolation Circulator Network," United States Patent No. 5,815,803, 1998.

[29] Ho, T. Q., "Matched Pair Circulator Antenna Isolation Circuit," United States Patent No. 6,313,713, 2001.

[30] Ackerman, E. I., and C. H. Cox. "Optimization of Photonic Transmit/Receive Module Performance," *Proc. of International Topical Meeting on Microwave Photonics,* 2009.

[31] Narashima, A., et al., "Tandem Single Sideband Modulation Scheme for Doubling Spectral Efficiency of Analogue Signals," *Electronics Letters,* Vol. 36, No. 13, 2000, pp. 1135–1136.

[32] Devgan, P., "Isolation of RF Signals Using Optical Single Side Band Modulation Combined with Optical Filtering," United States Patent No. 9,240,842, 2016.

[33] Schulze, H., and C. Lueders, *Theory and Applications of OFDM and CDMA: Wideband Wireless Communications,* Hoboken, NJ: John Wiley & Sons, 2005.

6

Signal Identification Utilizing RF Photonics

Over the previous chapters, the demonstration architecture has evolved to incorporate an RF circulator after the antenna, a remoting link for the antenna, and oscillators for signal generation. The next subsystem to be considered is one that allows for the identification of signals using RF photonics. A specific example is determining the center frequency of signals that are received at the antenna. As seen in Figure 6.1, the subsystem takes the RF signals from the antenna and then identifies the center frequencies. The information is then passed to the photonic filters and oscillators to assist with later processing of signals. Since the architecture is designed to have a very wideband capability, multiple signals will be received at the input of the system. Some of these signals will be of interest to the receiver, while others are meant to be ignored. In order to find the signals that are of interest, the center frequency of the various signals needs to be determined. The photonic filter can then be tuned to the center frequency in order to pass than signal and suppress the unwanted ones. In the case of the oscillator, it can be tuned to the center frequency for clock synchronization purposes. This chapter reviews methods for identifying the center frequency. In addition, the chapter presents a couple of methods for not only identifying the center frequency but also selectively amplifying chosen signals while suppressing other signals.

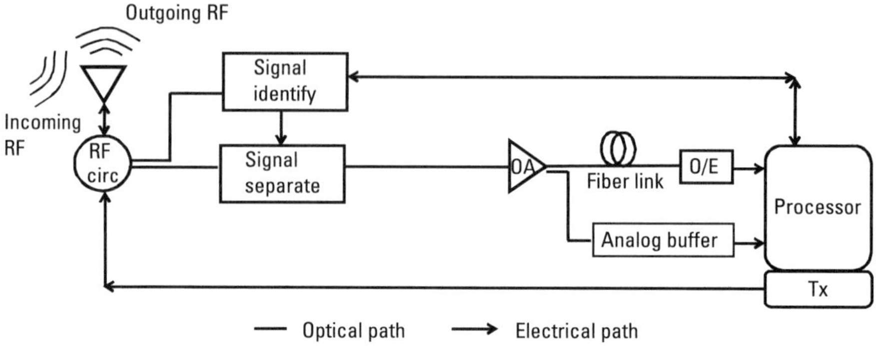

Figure 6.1 Demonstration architecture with signal identification block added.

6.1 Need for Signal Identification

The identification of the frequency of received RF signals is important in many fields, including radio astronomy. Radio astronomers often look for molecular and atomic spectral resonances in order to determine where celestial bodies are and what they consist of. The molecular frequencies cover a large range, including hydrogen at 1.4 GHz, methanol at 12.17 GHz, ammonia at 23.8 GHz, and carbon monoxide at 806 GHz [1]. Another application would be for a survey of the local RF environment. As mentioned in Chapter 5, the RF spectrum is becoming more and more crowded. Having the ability to identify what signals exist in the surrounding area can be useful for diagnosing interference as well as policing unauthorized use of protected frequency bands.

Ideally, a wideband frequency analyzer could be used to detect and identify the center frequencies of the received signals. In traditional electronic systems, a swept spectrum analyzer can be used for determining the frequency of the incoming signals [2]. Unfortunately, this electronic solution suffers from two main deficiencies: slow scanning speeds and limited bandwidths. As seen in Figure 6.2, the spectrum analyzer operates by using a tunable local oscillator connected to a mixer. The signals that pass through the mixer are downconverted to a frequency at the center of a bandpass filter. The filtered signal is then detected and displayed. The displayed frequency matches the center frequency of the tunable local oscillator. While the spectrum analyzer is fairly simple to implement, the use of a tunable local oscillator combined with a mixer for downconversion limits the frequency range of the spectrum analyzer. In addition, the slow tuning speed of the local oscillator can allow for signals to be missed. Photonic methods for determining the center

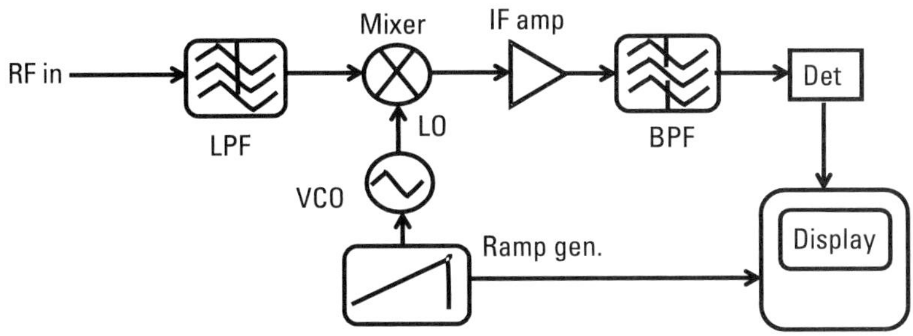

Figure 6.2 Block diagram of a scanning electrical spectrum analyzer (LPF: low-pass filter; LO: local oscillator; VCO: voltage-controlled oscillator; IF Amp: intermediate frequency amplifier; BPF: bandpass filter; and Det: envelope detector. (After: [3].)

frequency of the various signals coming into the system can overcome the electronic deficiencies.

6.2 Using RF Photonics for Spectrum Analysis

A photonic-based spectrum analyzer seeks to replace a traditional electronic spectrum analyzer by providing increased frequency range. One of the most straightforward implementations of a photonics-based spectrum analyzer is to replace the electronic components in the architecture shown in Figure 6.2 with photonic components [4]. As seen in Figure 6.3, the incoming RF signals are upconverted into the optical domain by use of an optical modulator, which takes the place of the mixer in the electronic implementation. The optical signal is then passed through a scanning Fabry-Perot (FP) filter, with the filter now scanning rather than the local oscillator in the electronic ver-

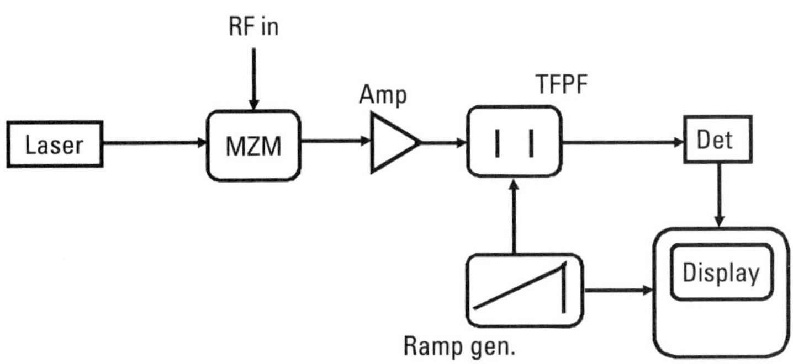

Figure 6.3 Block diagram of a scanning electrical spectrum analyzer (MZM: Mach-Zehnder modulator; Amp: optical amplifier; TFPF: tunable FP filter; and Det: optical photodetector). (After: [3].)

sion. The resulting output is then sent to a photodetector. As the FP filter is scanned, the optical sidebands due to the modulated RF signals will meet the resonance condition of the FP and pass through the filter with maximum transmission. If the scanning filter is synchronized to the output of the photodetector, a display of power as a function of RF frequency can be obtained. One of the advantages of the photonic architecture is that it does not require a downconverter, as the photonic components have more than enough bandwidth for the RF frequencies [4]. This makes the system scalable to higher frequencies without the need for additional electronics. However the system still has the disadvantage of being a scanning system and not instantaneously capturing wideband spectrum.

To overcome the disadvantage of scanning filter systems, another type of photonic spectrum analyzer has been investigated. Originally intended for use as a form of photonic-based memory, rare Earth-doped crystals can also be used as a spectrum analyzer [5]. Through a process known as spectral hole burning (SHB), the absorption of the crystal at a wavelength that is resonant with the atomic transitions can be modified. In one of the first demonstrations of this effect, an experimental setup was developed to show the feasibility of the spectrum analyzer. As seen in Figure 6.4, two laser beams arriving at different angles to the surface are used to write an absorption grating on the crystal. A third laser beam that contains the RF information is then input to the crystal from the opposite side of the first two beams. The absorption grating set up in the crystal by the first two beams then deflects the beam with the RF information onto a photodiode array. By precisely maintaining the power of the two optical beams that form the absorption grating, the deflection of the beam with the RF signals will be precisely mapped to the photodiode array. The results showed that the spectrum analyzer could resolve two different frequencies at a minimum resolution of 1.5 MHz [5]. While the original demonstration was limited by the photonic components that were used, a follow-on demonstration was able to improve the resolution to 500 kHz [6].

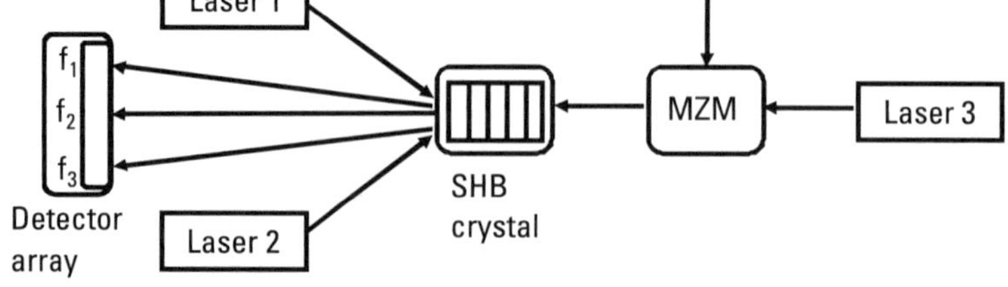

Figure 6.4 Spectrum analyzer based on a SHB crystal. Different RF frequencies are deflected onto different parts of the detector array. (After: [3].)

One of the advantages of this type of spectrum analyzer is the removal of a scanning filter. Since the bandwidth of the crystal is on the order of 100 GHz, the system can instantaneously identify the center frequencies of signals over that bandwidth, assuming that the RF signals can be modulated onto the optical beam before entering the crystal. Such a wide bandwidth operation would be very useful to radio astronomers [6]. Unfortunately the system has some disadvantages. The lifetime of excitation of the ions in the crystal is quite short and requires the crystal to be cooled on the order of 5° Kelvin in order to provide the minimum resolution. Also, the crystal has to have the grating constantly updated by the two optical beams. Finally the system operates at the resonance of the ions in the crystal, which in this case is 793 nm. Note that this does not match with the low-loss window of the optical fibers that are being used in the rest of the demonstration architecture. One possible solution is to use an optical frequency doubler in order to get the wavelength of the laser to match the crystal resonance. This does increase the complexity of the system, however. Nevertheless, these systems do show some promise to be able to perform spectrum analysis over a large instantaneous bandwidth.

6.3 Using Photonics Filters for Instantaneous Frequency Measurement

Other methods for determining the center frequency of RF signals have also been investigated. Most of them have focused on removing the scanning filter limitation and providing wide instantaneous bandwidth operation. One method for determining the center frequency involves the use of fixed optical filters. Unlike the scanning FP filter described in the previous section, the fixed optical filter has a sinusoidal response that can be combined with two lasers in order to determine the frequency of the RF signal [7]. The wavelength of one laser is set at the null of the sinusoidal response while the other laser is set at the peak of the filter response. This allows the generated sidebands to appear on complementary slopes of the filter response, as seen in Figure 6.5. The optical power of each pair of sidebands is separated and detected. A ratio of the powers from the two photodetectors, referred to as the amplitude comparison function (ACF), can then be used to determine the frequency of the RF signal. The transfer function of the filter for the two different laser wavelengths can be written as

$$H_{laser1}(\omega) = \left[1 + \cos\left(\frac{\omega_{RF}}{FSR}\right)\right] \tag{6.1a}$$

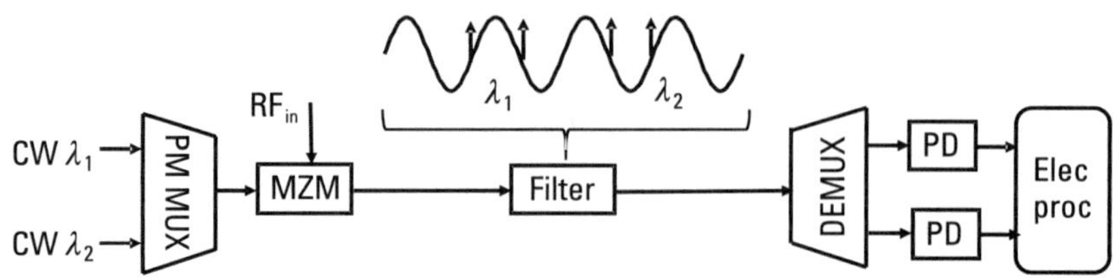

Figure 6.5 Two-laser system using an optical filter to identify RF center frequency, with the frequency response of the filter plotted above (PM MUX: polarization-maintaining wavelength multiplexer; DEMUX: wavelength demultiplexer; PD: photodiode; and Elec. Proc.: electronic processor. (After: [3].)

$$H_{laser2}(\omega) = \left[1 - \cos\left(\frac{\omega_{RF}}{FSR}\right)\right] \tag{6.1b}$$

where Ω_{RF} is the RF angular frequency and the FSR is the free spectral range of the filter. The ACF can be expressed as the ratio of the two transfer functions, which can then be described as

$$ACF = \left[\tan\left(\frac{\omega_{RF}}{2 * FSR}\right)\right]^2 \tag{6.2}$$

A plot of the ratio in Figure 6.6 shows that the ACF monotonically increases as a function of frequency. Thus, the frequency can be recovered by knowing the ACF response. An alternative approach, as shown in Figure 6.7, exchanges the two lasers for two filters [8]. The resulting response is equivalent to the two-laser case, with the advantage that the cost and power consumption of two filters is less than that of two lasers. The setups provide instantaneous measurement of the RF frequency with a good match to the theoretical performance. The disadvantage of these systems stems from their inability to measure more than one CW RF signal at a time. Also, these systems require the suppression of the optical carrier. If the optical carrier is not sufficiently reduced, it introduces a noise floor that limits the sensitivity of the overall system. Finally, the overall bandwidth of the system is limited to $f_{max} = FSR/2$ of the sinusoidal filter. Beyond this frequency, the system cannot disambiguate the frequency between the lower and upper half of the FSR of the filter. Despite these limitations, a similar system has been shown to measure the frequency of both a CW and pulsed RF signal at the same time [9]. Work continues to improve the performance of these systems.

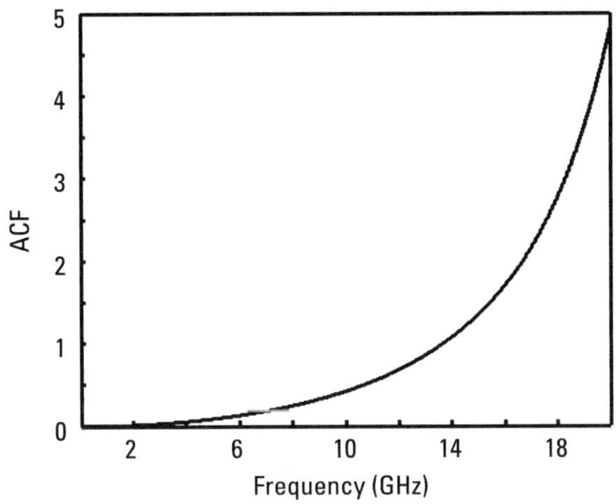

Figure 6.6 The ACF as a function of RF frequency for a two-laser system using an optical filter. (After: [3].)

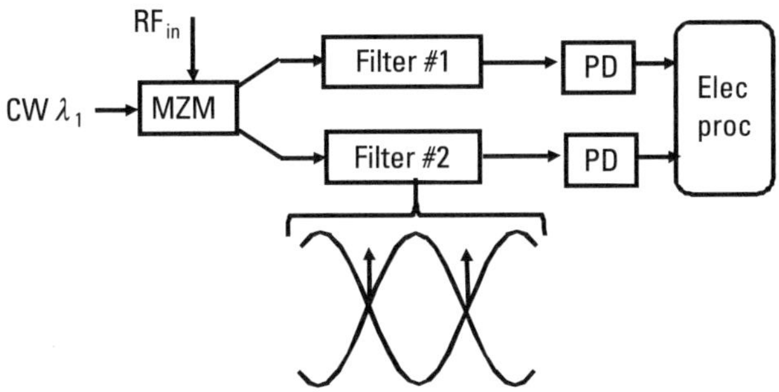

Figure 6.7 Two complementary filter systems for identifying the RF center frequency, with the frequency response of the two filters overlaid on each other (CW: CW laser source; PD. photodiode; and Elec. Proc.: electronic processor). (After: [3].)

6.4 Using Dispersion for Instantaneous Frequency Measurement

While using optical filters is one way to create the ACF response, another method is to use dispersion. As described in Chapter 2, dispersion in optical fibers leads to RF fading as a function of RF frequency when dual-sideband modulation is used. While often an unwanted response, the power fading can be seen as a narrow filter function. Making use of this function can provide an ACF similar to the one in the previous section. Since the dispersion is like a single filter that is wavelength-dependent, an initial demonstration was made with two lasers, as presented in Figure 6.8 [10]. In this case, the ACF will present a slope as shown in Figure 6.9 and described by

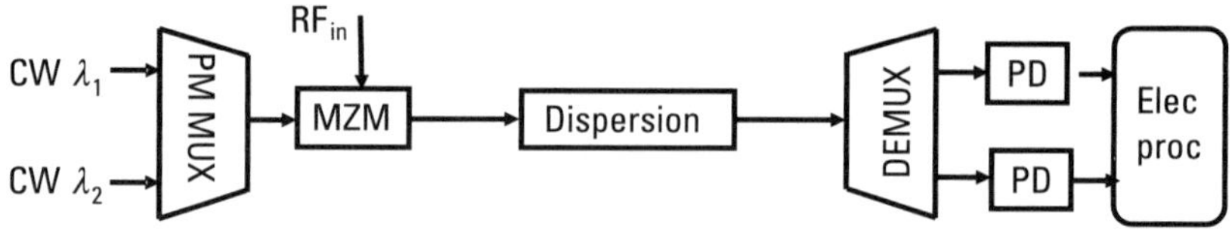

Figure 6.8 Two-laser system utilizing dispersion for identifying the RF center frequency (PM MUX: polarization maintaining wavelength multiplexer; DEMUX: wavelength demultiplexer; CW: CW laser source; PD: photodiode; and Elec. Proc.: electronic processor). (After: [3].)

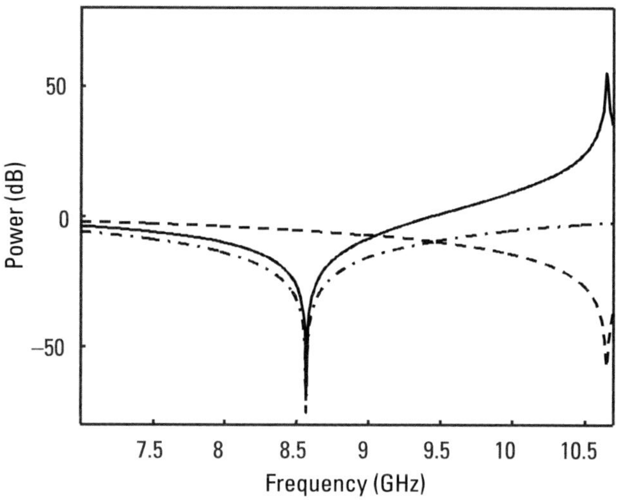

Figure 6.9 Power as a function of RF frequency for the two different wavelengths (dashed lines) after passing through the dispersive medium. The ratio of the responses is shown in the solid line. (After: [3].)

$$ACF = \frac{\cos^2\left(\pi DL\lambda_1^2 f^2/c\right)}{\cos^2\left(\pi DL\lambda_2^2 f^2/c\right)} \tag{6.3}$$

where D is the dispersion (ps/(nm•km)), L is the length of the dispersive medium, c is the speed of light, f is the RF frequency, and $\lambda_{1,2}$ is the wavelength of the two lasers, respectively.

The two dashed lines are the RF response for each of the two lasers individually. By taking the ratio of the responses in the electrical processing unit, the solid black line ACF can be generated. The advantage of this system is the high slope of the ACF, which allows for a small change in frequency to result in a large change in amplitude. The disadvantage is that the system still cannot recover the frequency of more than one signal at a time. In addition, the demonstration was limited to a frequency range of 7–10.5 GHz for the chosen dispersion of each laser.

As an improvement, another version of the dispersion-based system was developed to be able to measure simultaneous signals [11]. Figure 6.10 shows an example system with a single laser followed by a dispersive element and a photodetector. In this case, the dispersion is actually used to perform a frequency-to-time mapping of the RF-generated dual optical sidebands. The system takes advantage of the fact that the time delay through a dispersive medium can have a linear response as a function of wavelength. Using this capability, each optical sideband will arrive at the photodetector at a different time, as seen in Figure 6.10.The difference in time between the arrival of the first sideband and its corresponding twin is directly proportional to twice the RF frequency. Therefore, measuring the time delay and knowing the slope of the time delay through the dispersive medium will yield the corresponding RF frequencies. In the demonstration, two RF signals of 20 and 40 GHz were detected and the frequency recovered through the frequency-to-time mapping process [11]. However, the demonstration was limited to frequency separations of 12.5 GHz. The advantage of this system is that multiple RF signals can be modulated onto the optical carrier of the laser, and the frequencies can

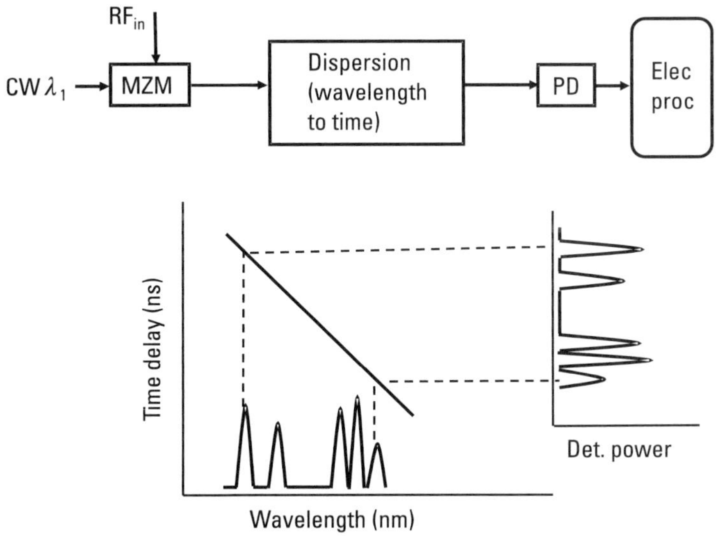

Figure 6.10 Single-laser system using dispersion to map frequency to time. Taking the received optical power at the photodiode and measuring the time difference between changes will yield the frequency of the RF signals. (After: [3].)

be detected. The system is limited by the resolution of the timing system used to differentiate between signals, as well as any deviation from linear that the dispersion slope may have. Regardless, this time-based solution offers capabilities beyond some of the other systems previously described.

6.5 Combinations of Different Methods for Frequency Measurement

A combination of the scanning local oscillator and dispersion has also been investigated for IFM. These systems can identify the center frequencies of multiple RF signals [12, 13]. Again, making use of the power fading as a function of frequency in a dispersive medium, a tunable laser can be used to sweep through the dispersive medium. The result can be thought of as a scanning filter as seen in Figure 6.11. By tracking the tuning laser wavelength, the power measurement can be translated into a frequency and power measurement of the RF signals using a digital processor. In the first demonstration [12], the ability to successfully resolve two different frequencies at least 15 MHz apart was shown. The measured power of the RF signal was also shown to be within 0.1 dB of the actual power. Still, the maximum measured frequency was limited to 18 GHz due to photonic components used. Another

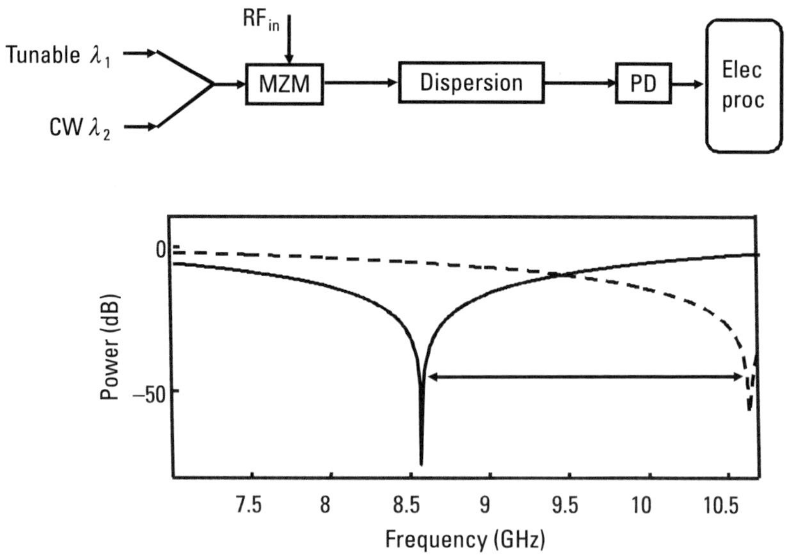

Figure 6.11 Using a tunable laser along with a fixed laser to move the RF power notch as a function of frequency. (After: [3].)

demonstration was shown to expand the frequency operation to 24 GHz while also demonstrating a sensitivity of –48 dBm [13].

6.6 Using FIR and IIR Filters for Instantaneous Frequency Measurement

The previous demonstrations have been interested in using optical filters, dispersion, or some combination of the two in order to recover the center frequency of the RF signals. Another method to determine the frequency is to use either FIR or IIR filters. Looking back at Chapter 5, FIR and IIR filters can be realized by using photonic means. In one demonstration, a combination of FIR and IIR filters can be used to identify the center frequency of an RF signal [14]. The system is presented in Figure 6.12, where the FIR filter is generated using photonic means, and the IIR filter is an electronic implementation. Note that the gain in the overall loop is set below threshold for oscillation to prevent the cavity from becoming an RF source. The combination of the two filter responses yields a transfer response of

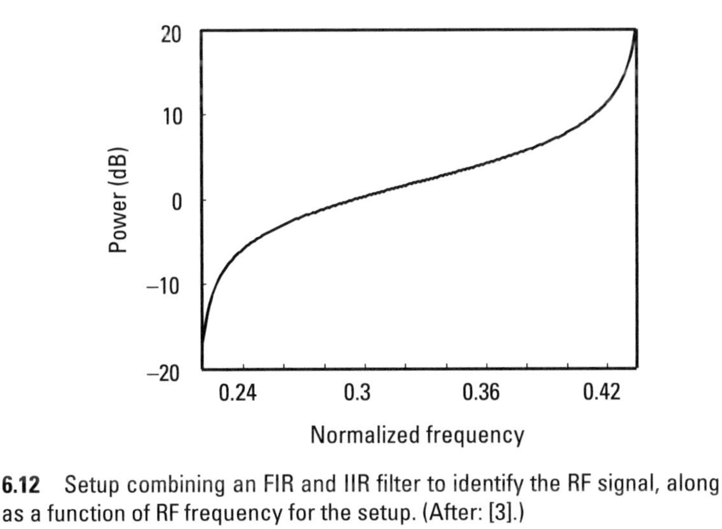

Figure 6.12 Setup combining an FIR and IIR filter to identify the RF signal, along with the power as a function of RF frequency for the setup. (After: [3].)

$$H(\omega) = \frac{L_{feedback}\left(1 + e^{j\omega_{RF}T}\right)}{e^{j\omega_{RF}T} - G/2} \tag{6.4}$$

where Ω_{RF} is the RF angular frequency, L_{feedback} is the loss in the opto-electronic feedback loop, G is the gain in the loop, and T is the time delay in the loop. Note that the time delay of the IIR feedback is set to match the time delay of the two arms of the FIR filter [14]. For a given set of parameters the detected power increases as a function of frequency, which is shown in Figure 6.12. The response is similar to demonstrations using dispersive methods. The FSR, as well as the Q factor, of the overall filter response is limited by the long length of lossy RF cable used in the system. In order to remove the limitation of the IIR filter, a photonic implementation was used [15]. As discussed in Chapter 5, the photonic IIR filter implementation requires a broadband light source in order to remove any coherence effects. The long RF cable was replaced with an optical fiber loop with an optical amplifier. While the loss of the optical fiber is much lower than the RF cable, the length was not much shorter. Therefore, the FSR was still shown around 42 MHz. Again, the gain in the optical loop was set below threshold to make sure there would be no oscillatory behavior. Still, the input power of the RF signals is limited to a range between –7 and –16 dBm, severely reducing the dynamic range of the measurement system.

6.7 Frequency Measurement with Multimode Photonic Systems

As mentioned previously, the use of a fiber loop can provide advantages in terms of the filter response with a much smaller form factor. The dispersion demonstrations require long lengths of fiber or specialty devices to give the proper response. The fiber loop can provide a similar response with a much smaller length of fiber. Yet, the previous demonstrations were limited in the sensitivity to the RF signal power at the input of the system. In order to address these deficiencies, a different photonic system can be used. Specifically, a multimode OEO can provide enhanced sensitivity as well as selectivity.

As seen in previous chapters, the OEO can be used as a low-noise oscillator that can generate RF signals. However, the OEO does have applications in analog systems beyond the ones described previously. One such application is the identification of low-power RF signals in a cluttered environment. Imagine a wide frequency spectrum containing multiple RF signals, all with varying powers as shown in Figure 6.13. In order to identify a subset of the RF signals from the others, signals of interest need to be amplified while

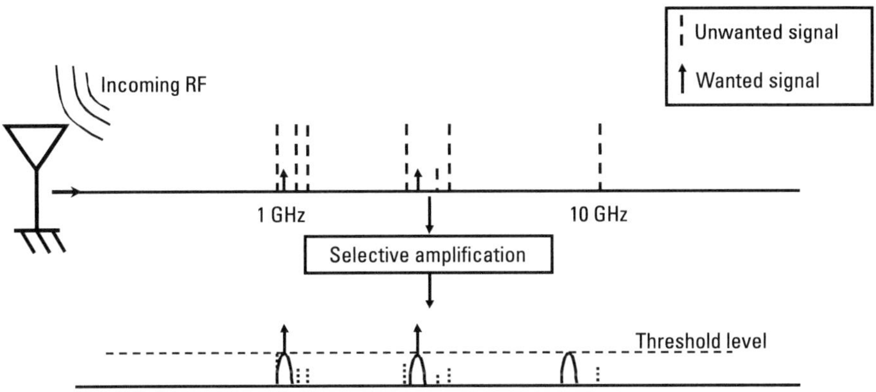

Figure 6.13 RF signals covering 1–10 GHz of bandwidth from an antenna before and after selective amplification.

suppressing the others below some chosen threshold level. A narrowband electronic filter combined with an RF amplifier could accomplish this task. Such a system is hard to realize at higher frequencies, is not widely tunable, and has limited sensitivity to low-power signals. An OEO however can handle this task. Specifically, the injection locking process in an OEO can be used to amplify certain RF signals while suppressing the other ones. Injection locking on an OEO occurs when an external RF signal is injected into the OEO cavity, as seen in previous demonstrations [16]. The frequency of the signal generated by the OEO will become locked to the frequency of the injected RF signal. In a traditional OEO, the range of frequencies to which the OEO can be locked is limited by the electronic bandpass filter used inside the OEO cavity. For the case of signal identification and amplification, the electronic bandpass filter is removed from the OEO cavity, and the power inside the OEO is set just below threshold. The setup is shown in Figure 6.14. Such an OEO is referred to as a multimode OEO (MM-OEO). When an external signal is injected into the OEO cavity, two results can occur. If the injected signal's frequency matches one of the cavity modes of the MM-OEO, it will cause the OEO cavity to oscillate, and the signal will see gain. If the injected signal's frequency does not match one of the cavity modes, the signal will see

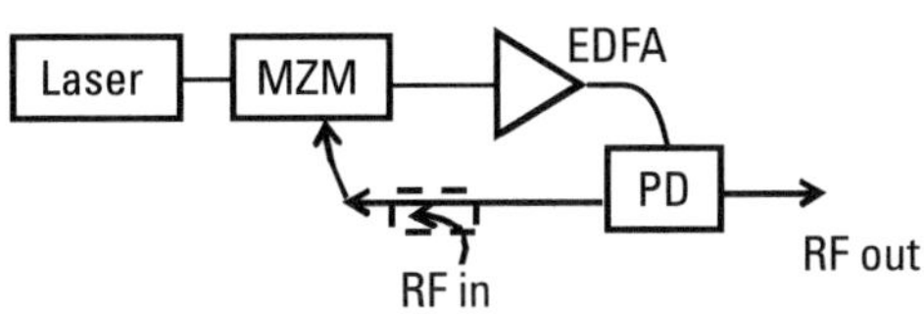

Figure 6.14 Setup of a MM-OEO. (After: [3].)

loss and be suppressed. This process is depicted in the movie at the website found on page *iv* of this book, where an injected signal is swept from one mode to an adjacent one, and the resulting gain and loss is observed. The first demonstration of an MM-OEO used an electronic RF amplifier to show the loss and gain profile over multiple cavity modes [17]. Because the RF amplifier adds noise to the system, the work was extended to an MM-OEO that used all-photonic gain [18]. The all-photonic gain uses the method of a low-biased MZM followed by an optical amplifier, as discussed in the previous chapters. Due to the long length of fiber in the EDFA used for the all-photonic gain, the MM-OEO cavity modes are spaced about 5.2 MHz apart, with each cavity mode having a gain bandwidth of 1 MHz. The MM-OEO has been shown to provide gain for RF signals at frequencies as high as 6 GHz. In order to characterize the MM-OEO further, the sensitivity and compression dynamic range are measured. The sensitivity of the MM-OEO is as low as -83 dBm, while the maximum input power into the MM-OEO before the gain becomes compressed is –11 dBm. This yields a compression dynamic range of 72 dB. In order to select a small subset of cavity modes from the MM-OEO, the optical output of the MM-OEO was passed through an integrated silicon FP filter (FPF). The FPF has a 3-dB bandwidth of 4.5 GHz, and after passing the optical signal from the MM-OEO through the FPF, the filtered RF gain is only positive at frequencies of 3 GHz or lower, as opposed to 6 GHz without the FPF. Thus, an optical filter can be used to identify the RF frequency of the MM-OEO. The advantage of the integrated silicon FPF is the ability to fabricate up to 100 filters on a single chip. This scalability in a small size can be advantageous in systems wanting to identify RF signals of multiple gigahertz in bandwidth.

While the MM-OEO described above shows promise for use in selective amplification of RF signals, it is desirable to increase the mode spacing of the MM-OEO cavity, which requires shortening the length of the cavity. As mentioned previously, the EDFA adds an equivalent of approximately 10m of fiber into the cavity. In order to overcome this limitation while still using all-photonic gain in the MM-OEO cavity, the EDFA was replaced with a semiconductor optical amplifier (SOA). This device uses a semiconductor gain medium that can be on the order of a centimeter in length, plus the fiber pigtails in order to connect it to the other fiber components in the MM-OEO. Such a length savings can significantly improve the cavity mode spacing. In this new configuration, an MM-OEO was demonstrated having a mode spacing of 100 MHz, which is a 20-times improvement over the previous demonstration [19]. While the mode gain bandwidth increases also from 1 to 11 MHz, the overall selectivity of this MM-OEO is better than the previous demonstration. Another advantage of increasing the mode spacing is that it allows the MM-OEO to be used with two lasers. By proper selection of the DC bias of the MZM, the two lasers will allow the MM-OEO to switch between complementary modes in the MM-OEO cavity. The modes are complementary in that in one condition, a given frequency will see gain (loss), while in the other condition, the same frequency will now see loss (gain). As seen in Figure 6.15, with one laser on, one set of modes have gain in the MM-OEO. However when both lasers are on, the complementary set of cavity modes now see gain. This allows the MM-OEO to selectively amplify more frequencies with the same cavity than is possible with just one laser. Another improvement in the system comes from the sensitivity of this MM-OEO. With the increased mode spacing, the integrated noise of the MM-OEO can be measured over 140 MHz while only incorporating one cavity mode. With a measured gain of 10 dB, input sensitivity is calculated to be –78.4 dBm. When compared to thermal noise over the same 140-MHz bandwidth (–92.6 dBm), the MM-OEO is only 14.2 dB above the thermal limit. Thus the MM-OEO can be used to amplify very low-power RF signals for detection.

The MM-OEO has been shown to offer a method for selectively amplifying low-power RF signals. The system can have very good input sensitivity, high CDR, and large selectivity. Further improvements of the MM-OEO can be made, the first being to further shorten the cavity in order to get the mode spacing to 1 GHz, while increasing the overall bandwidth to 10 GHz and higher. Such improvements can make the MM-OEO attractive

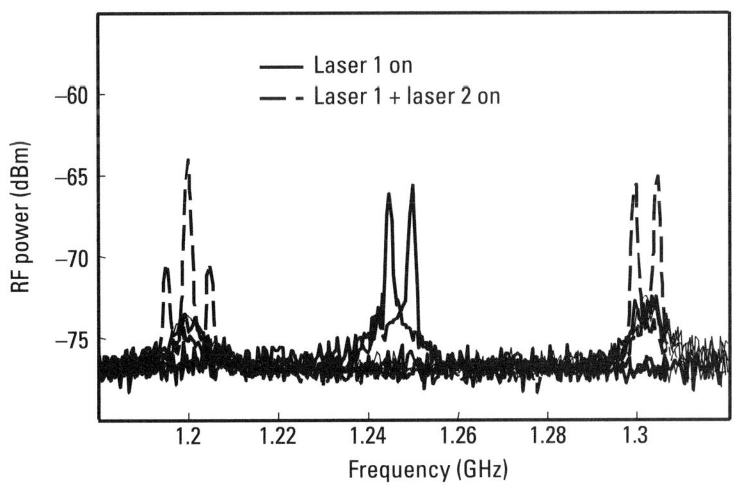

Figure 6.15 The two-laser MM-OEO showing selective amplification of complementary modes when either one or both lasers are on. (After: [20].)

to applications such as identification of multiple RF signals in a cluttered environment.

6.8 RF Frequency Identification Using Optical Injection Locking

While the MM-OEO has shown promise in terms of being able to select out and identify signals over a wide range of RF frequencies it still is limited to a fairly closely spaced set of modes. In addition, the suppression of unwanted signals was only 25 dB compared to the signal of interest. Finally the tuning of the modes of the cavity is achieved by changing the length of the OEO cavity. While using a fiber stretcher can tune the modes of the OEO cavity, the speed of tuning is limited. Another method for amplifying signals involves the use of optical injection locking of two separate lasers [21]. Resembling the optical injection locking process for making a fast tunable oscillator, the two-laser system can be used to discriminate and identify RF signals. The system works by amplifying the selected RF signal while completely suppressing all other RF signals, regardless of their power. By injecting the output of a master laser that has been phase-modulated with RF signals into a slave laser, selected RF signals will see gain while completely suppressing other unwanted signals. The system can be frequency-tuned very quickly with very high extinction ratio between gain and loss. The system is different than other injection locking methods [22] in a couple of different ways: First, the slave laser is multimode, and second, the slave laser is kept below threshold in order to give gain to the appropriate sideband. Also, this system is designed for fast-tunable, RF signal discrimination, which is different from other demonstrations. Finally, most demonstrations use a MZM instead of a phase modulator.

The system for RF signal discrimination is illustrated in Figure 6.16. The optical output of the master laser connects to the optical input of a phase modulator. The RF input of the phase modulator receives the signals, which are illustrated to demonstrate the different operation. The optical output of the phase modulator is injected into the optical output of the multimode slave laser via an optical circulator (port 1-2). The mixed signal generated inside the slave laser is output to an optical filter in order to select one of the modes from the slave laser via the third port of the optical circulator. The filtered optical signal is connected to the input of a photodetector. The resulting RF signal from the photodetector is measured. The drive current of the multimode slave laser is set below threshold in order to provide gain to the incoming RF-modulated signal from the master laser. In one case, the wavelength of the master laser is set to the wavelength of one of the modes of the slave laser minus the RF frequency of interest. Under this condition, the

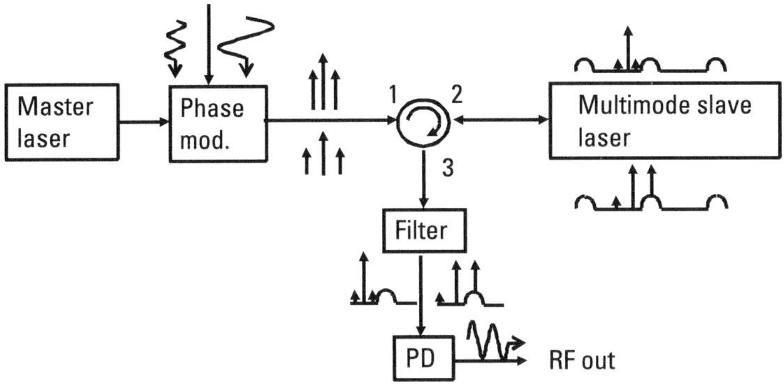

Figure 6.16 Optical injection locking method for selectively amplifying RF signals along with identifying them. (After: [21].)

wavelength of the upper sideband generated by the RF signal on the master laser will match the wavelength of one of the modes of the slave laser, while the lower sideband does not match a nonoscillating mode of the slave laser. In this case, the upper sideband will see gain while the lower sideband will see loss. After the photodetector, the resulting RF signal will see gain as compared to the RF signal at the phase modulator input. Note that the system can be set to amplify the lower sideband with the same result. In the other case, when the RF signal sidebands do not match the wavelength of any of the modes of the slave laser, there will be no RF signal present at the output of the photodetector. This is inherent to the phase modulation process as long as the sidebands' power levels are equal. Thus, the system will completely suppress the RF signal if it does not match the appropriate conditions.

When the optical sideband is set to the wavelength of one of the modes of the slave laser, the resulting sideband is higher than the other sideband, which sees optical loss. This allows the otherwise balanced signal to be detected at the photodetector. A measurement has been made with an RF signal at 14.18 GHz input to the phase modulator. The input power is –40 dBm, and the measured output power is –39 dBm, providing a gain of 1 dB. When the RF signal frequency is moved off of the slave laser's operating wavelength, there is no observable signal. In the optical spectrum, the side-mode is amplified to the same level as the carrier. When the signal is moved off the operating point, the sideband is much lower than the optical carrier. At the point that the sideband is not amplified, there is no observable signal at the output of the photodetector. The system will also detect RF signals that are modulated by other data. For example, the RF signal at 14.18 GHz has been modulated with a 500-kHz pulse. The detected RF signal is amplified along with the 500-kHz sidebands, allowing the data to be recovered.

The additional sidebands are higher frequency components due to the pulse shape. Note that any modulation can be amplified as long as it is within the gain bandwidth of the system. By changing the bias current of the slave laser, the wavelength changes, consequently changing the RF frequency that will be amplified. In this case, the current can be changed on the order of ~100 nsec. The RF frequency to be amplified can be tuned over a frequency range of > 1 GHz in that time frame.

A mathematical description of the phase modulation is shown in (6.5). The phase modulator has a transfer function given by

$$E_{out}(t) = \sqrt{\alpha_{PM} G_{opt}}\, e^{i\varphi(t)} E_{in}(t) \tag{6.5}$$

where α_{PM} is the optical insertion loss of the phase modulator, G_{opt} is optical gain/loss before the photodetector, $\varphi(t) = \varphi_{rf}\sin(\Omega_{rf}t)$ and $\varphi_{rf} = \pi(V_{rf}/V_{\pi,rf}(\Omega_{rf}))$. Now using the Jacobi Anger expansion $e^{iz\sin\theta} = \sum_{n=-\infty}^{\infty} J_n(z)e^{in\theta}$ the following can be written:

$$\begin{aligned}
E_{carrier}(t) &= \sqrt{\alpha_{PM} G_{opt}}\, \overline{E}_{in} e^{i\omega_o t}\left[J_0(\varphi_{rf})\right], \\
E_{usb,fund}(t) &= \sqrt{\alpha_{PM} G_{opt}}\, \overline{E}_{in} e^{i\omega_o t - i\Omega_{rf} t}\left[-J_1(\varphi_{rf})\right], \\
E_{lsb,fund}(t) &= \sqrt{\alpha_{PM} G_{opt}}\, \overline{E}_{in} e^{i\omega_o t + i\Omega_{rf} t}\left[J_1(\varphi_{rf})\right],
\end{aligned} \tag{6.6}$$

where Ω_o is the optical frequency of the laser and J_n is the Bessel function of the *n*th order. At the photodetector, the carrier field beats with both the upper and lower sidebands in order to generate an optical photocurrent. In the case of phase modulation, the resulting mixing terms are given below:

$$\begin{aligned}
E_{carrier} \times E^*{}_{usb,fund} &= \alpha_{PM} P_{laser} G_{opt}\left[J_0(\varphi_{rf})\right]\left[-J_1(\varphi_{rf})\right]e^{i\Omega_{rf} t}, \\
E^*{}_{carrier} \times E_{usb,fund} &= \alpha_{PM} P_{laser} G_{opt}\left[J_0(\varphi_{rf})\right]\left[-J_1(\varphi_{rf})\right]e^{-i\Omega_{rf} t}, \\
E_{carrier} \times E^*{}_{lsb,fund} &= \alpha_{PM} P_{laser} G_{opt}\left[J_0(\varphi_{rf})\right]\left[J_1(\varphi_{rf})\right]e^{-i\Omega_{rf} t}, \\
E^*{}_{carrier} \times E_{lsb,fund} &= \alpha_{PM} P_{laser} G_{opt}\left[J_0(\varphi_{rf})\right]\left[J_1(\varphi_{rf})\right]e^{i\Omega_{rf} t},
\end{aligned} \tag{6.7}$$

where $E_{in}E_{in}^* = P_{laser}$. By inspection, when the four terms are added together they cancel and no RF photocurrent will be generated. So as long as the two sidebands are equal in power, the RF signal will be completely suppressed at the output of the photodetector. However, when the phase-modulated optical signal is injection-locked into the slave laser, one sideband can be made higher than the other sideband. For this case we make the upper sideband larger than the lower sideband. We can represent that by making G_{opt} different for the upper sideband than for the lower sideband. Including this in (6.7), we now determine that the terms do not add to zero but give the following result for the photocurrent:

$$I_{PD} = R\alpha_{PM} P_{laser}\left({G_{opt,usb}}/{G_{opt,lsb}} - 1 \right) J_0(\varphi_{rf}) J_1(\varphi_{rf}) \cos\left(\Omega_{rf} t\right), \quad (6.8)$$

where there is a different G_{opt} for the lower and upper sideband and R is the responsivity of the photodetector. Since we are interested in low-power RF signals, we can take the small signal approximation for the Bessel functions and find the RMS output RF power given by $P_{rf} = 1/2 I_{PD}^2 Z_{out}$, with Z_{out} is the output impedance of 50Ω. The small signal RF power is then given by the following:

$$P_{rf,ss,out} = \frac{1}{4}\left(R\alpha_{PM} P_{laser}\right)^2 \left({G_{opt,usb}}/{G_{opt,lsb}} - 1 \right)^2 \pi^2 \frac{V_{rf}^2}{V_{\pi,rf}^2} \varphi_{rf} Z_{out}. \quad (6.9)$$

Now the RF gain is simply the ratio of output RF power to the input RF power, which is given by $P_{rf,in} = V_{rf}/(2Z_{in})$ with Z_{in} as the input impedance. The RF gain is given by the following in both linear and decibel units.

$$\begin{aligned} G_{rf} &= \frac{\left(R\alpha_{PM} P_{laser}\right)^2 \left({G_{opt,usb}}/{G_{opt,lsb}} - 1 \right)^2 \pi^2 Z_{out} Z_{in}}{4V_{\pi,rf}^2}, \text{(linear)} \\ G_{rf} &= -22.1 + 20\log_{10}(\Re\alpha_{PM} P_{laser}) - 20\log_{10}(V_{\pi,rf}) \\ &+20\log_{10}({G_{opt,usb}}/{G_{opt,lsb}} - 1), (dB). \end{aligned} \quad (6.10)$$

The system can see positive gain when the ratio of sidebands is enough to compensate for the $V_{\pi he}$ of the modulator. This mathematical description demonstrates the use of the system to either completely suppress an RF signal or to provide gain to an RF signal.

This method provides an RF identifier with fast tuning speeds that can provide gain to specific RF signals while completely suppressing other RF signals. The fast tuning speed is useful for reconfigurable systems that need to change the RF discriminator's frequency very quickly. The use of injection-locked lasers improves the mode spacing when compared to MM-OEO demonstrations. The ability to provide gain without the use of an electronic amplifier increases the frequency range over which the system works when compared to systems that require electronic gain. The system is also more compact when compared to the MM-OEO demonstration.

6.9 Conclusions

A review of photonic methods for determining the center frequency of RF signals entering the demonstration architecture has been performed. The first method is a photonic-based spectrum analyzer. While a scanning photonic analyzer has wider bandwidths and does not require a downconverter, it still suffers from relatively slow sweeping speeds. A more advanced photonic spectrum analyzer, based on SHB in a rare Earth-doped crystal, can instantaneously capture the entire frequency band without sweeping. However, the crystal requires constant updates from the writing laser and needs to be cooled to 5K. Beyond spectrum analyzer-based methods, the use of optical filters has been used to identify the frequency of an RF signal. While initial demonstrations were limited to only one RF signal, more recent methods have shown the capability to identify the frequency of multiple RF signals, including ones that are continuous as well as pulsed. Other methods have used the frequency-dependent loss as a function of RF frequency in a dispersive medium to identify the RF signal. Another subset uses a combination of photonic-based FIR and IIR filtering responses to identify the frequency of the signal. Finally, the use of multimode photonic systems can be used to identify the frequency, while suppressing unwanted signals and offering the ability to be tuned very quickly. Most, if not all of the methods, work by using a transfer function to convert the frequency of the incoming RF signal into a change in optical power. By making a unique transfer function, the signal's frequencies can be identified. The identification of the frequency of the incoming signal will be important for Chapter 7, where signal-processing techniques using photonics are introduced. Many of these require some knowledge of the center frequency in order to operate on the signals.

References

[1] Keysight Technologies, Spectrum Analysis Basics, Application Note 150, www.keysight.com.

[2] http://www.setileague.org/articles/protectd.htm.

[3] Devgan, P., "Signal Identification and Isolation Utilizing Radio Frequency Photonics," AFRL Report, AFRL-RY-WP-TR-2017-0158, 2017.

[4] Winnall, S. T., and A. C. Lindsay, "A Fabry-Perot Scanning Receiver For Microwave Signal Processing," *Transactions on Microwave Theory and Techniques,* Vol. 47, No. 7, 1999, pp. 1385–1390.

[5] Ménager, L., et al., "Demonstration of a Radio-Frequency Spectrum Analyzer Based on Spectral Hole Burning," *Optics Letters,* Vol. 26, No. 16, 2001, pp. 1245–1247.

[6] Lavielle, V., et al., "Wideband Versatile Radio-Frequency Spectrum Analyzer," *Optics Letters,* Vol. 28, No. 6, 2003, pp. 384–386.

[7] Chi, H., X. Zou and J. Yao, "An Approach to the Measurement of Microwave Frequency Based on Optical Power Monitoring," *Photonics Technology Letters,* Vol. 20, No. 14, 2008, pp.1249–1251.

[8] Zou, X., H. Chi and J. Yao, "Microwave frequency Measurement Based on Optical Power Monitoring Using a Complementary Optical Filter Pair," *Transactions on Microwave Theory and Techniques,* Vol. 57, No. 2,2009, pp. 505–511.

[9] Lu, B., et al., "Photonic Frequency Measurement and Signal Separation for Pulsed/CW Microwave Signal," *Photonics Technology Letters,* Vol. 25, No. 5, 2013, pp. 500–503.

[10] Nyugyen, L. and D. Hunter, "A Photonic Technique for Microwave Frequency Measurement," *Photonics Technology Letters,* Vol. 18, No. 10, 2006, pp. 1188–1190.

[11] Nyugyen L., "Microwave Photonic Technique for Frequency Measurement of Simultaneous Signals," *Photonics Technology Letters,* Vol. 21, No. 10, 2009, pp. 642–644.

[12] Vidal, B., T. Mengual, and J. Marti, "Photonic Technique for the Measurement of Frequency and Power of Multiple Microwave Signals," *Transactions on Microwave Theory and Techniques,* Vol. 58, No. 11, 2010, pp. 3103–3108.

[13] Wang, Y., et al., "Photonic Approach for Microwave Spectral Analysis Based on Fourier Cosine Transform," *Optics Letters,* Vol. 36, No. 19, 2011, pp. 3897–3899.

[14] Zhou, J., et al., "Instantaneous Microwave Frequency Measurement Using a Photonic Microwave Filter with an Infinite Impulse Response," *Photonics Technology Letters,* Vol. 22, No. 10, 2010, pp. 682–684.

[15] Niu, J., et al., "Instantaneous Microwave Frequency Measurement Based on Amplified Fiber-Optic Recirculating Delay Loop and Broadband Incoherent Light Source," *Journal of Lightwave Technology,* Vol. 29, No. 1, 2011, pp. 78–84.

[16] Lasri, J., et al., "Ultra-Low Timing Jitter 40Gb/s Clock Recovery Using a Self-Starting Optoelectronic Oscillator," *Photonics Technology Letters,* Vol.16, No. 1, 2004, pp. 263–265.

[17] Urick, V. J., et al., "Channelisation of Radio-Frequency Signals Using Optoelectronic Oscillator," *Electronics Letters,* Vol. 45, No. 24, 2009, pp. 1242–1244.

[18] Devgan, P. S., et al., "Detecting Low-Power RF Signals Using a Multimode Optoelectronic Oscillator and Integrated Optical Filter," *Photonics Technology Letters,* Vol. 22, No. 3, 2010, pp. 152–154.

[19] Devgan, P. S., V. J. Urick, and K. J. Williams, "Detection of Low-Power RF Signals Using a Two Laser Multimode Optoelectronic Oscillator," *Photonics Technology Letters,* Vol. 24, No. 10, 2012, pp. 857–859.

[20] Devgan, P., V. J. Urick, and K. J. Williams, "Multimode Optoelectronic Oscillator," United States Patent 8,824,901, 2014.

[21] Devgan P. and N. Usechak, "Discrimination of Low Power RF Signals Using Phase Modulation with Injection Locked Lasers," United States Patent 9,435,839, 2016.

[22] Hong, C., et al., "Single-Sideband Modulation Based on an Injection-Locked DFB Laser in Radio-Over-Fiber Systems," *Photonics Technology Letters,* Vol.22, No. 7, 2010, pp. 462–464.

7

Signal Processing Utilizing RF Photonics

Up to this point, the demonstration architecture has been incorporating different subsystems linking the antenna to the signal processing block. These subsystems are designed to identify and separate specific signals for further processing. Often signal processing is done in the digital domain, where the advantages provided by high-speed computing can be leveraged. However, the limitations of the bandwidth of digital systems still leave a need for analog signal-processing applications. Analog signal-processing systems are often used to prepare signals for later digital analysis. Examples of analog signal processing include downconversion, phase corrections, and even analog-to-digital conversion. All of these applications can be accomplished by RF photonic means. Along with the signal-processing block, the demonstration architecture also contains a transmit block that is connected to the antenna. Transmission often requires the generation of unique, wideband waveforms. RF photonic methods for generating these types of waveforms can be useful for the transmission block. In Figure 7.1, the processing and transmission blocks of the demonstration architecture are expended to show a frequency downconverter, an ADC, and an arbitrary waveform generator. This chapter reviews downconversion of chosen RF signals. In addition, the chapter covers analog-to-digital conversion and waveform generation parts.

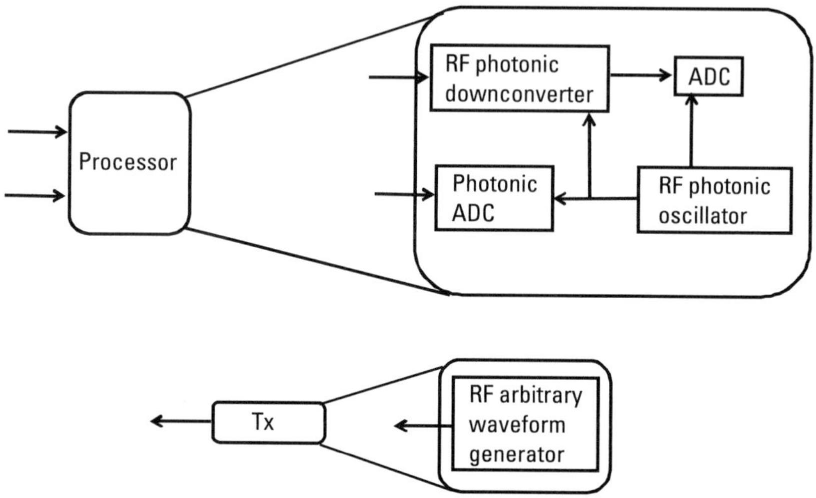

Figure 7.1 Demonstration architecture processing and transmit blocks, expanded view.

7.1 Need for Downconversion

Most traditional RF systems require a downconversion stage before the signals can be digitized. As mentioned previously, the SNR of the ADC is directly related to the ENOB it can produce. For example, an ADC with an ENOB of 8 has an SNR of 50 dB (see [4.1b]). The SNR of the ADC will increase by 6 dB for every extra bit. However, as the input analog bandwidth of the ADC increases, the ENOB decreases, due to various limitations that include quantization error as well as timing jitter on the sampling clock. In order to digitize the RF signals with the highest fidelity, a balance must be struck between the ENOB and analog bandwidth of an ADC. For most wideband applications, an ADC bandwidth of 500 MHz can operate with an ENOB of around 10. Note that a bandwidth of 500 MHz requires a sampling rate of at least 1 GHz in order to satisfy the Nyquist relation. Still, the center frequency of the RF signals received by the wideband system can be much higher than the ADC bandwidth. Thus, a downconverter is needed to shift the frequency into the operational range of the ADC.

An RF downconverter consists of a mixer, along with amplification and filtering. An example appears in Figure 7.2. The mixer is a three-port device that takes in the RF signal at the input, mixes it with a local oscillator (LO)

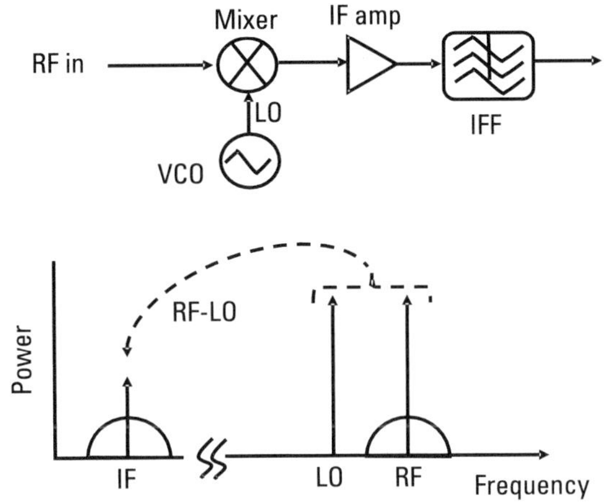

Figure 7.2 Block diagram of a downconverter (IFF: intermediate frequency filter; LO: local oscillator; VCO: voltage-controlled oscillator; and IF Amp: intermediate frequency amplifier). (After: [1].)

and produces an intermediate frequency (IF) at the output. The IF output is determined by the frequency of the RF signal and the LO. For example, an RF of 10 GHz will mix with an LO of 9.5 GHz to generate IF outputs at 19.5 GHz (sum frequency) and 0.5 GHz (difference frequency), along with higher-order mixing terms. The IF is usually limited to match the analog bandwidth of the ADC by using low-pass filters that only allow the IF of interest to pass while suppressing the unwanted higher-order mixing terms. The mixing process also introduces loss that is measured as the conversion efficiency. To compensate for a conversion efficiency less than 100%, amplifiers are used to increase the power of the IF that appears at the output. Since the downconverter uses a mixer and an amplifier, it has a noise figure and SFDR that are often specified by the manufacturer. In addition, the phase noise of the LO can be mixed with the incoming RF signal which will degrade the generated IF. While the downconverter does not supply the LO, it does specify the maximum phase noise allowed at the LO input before the IF performance is degraded beyond the intrinsic noise of the downconverter. One additional performance metric of the downconverter is the isolation. While the downconverter would ideally allow neither the LO nor RF signal to leak out the IF output, in practice either of the signals can appear. The isolation is a measure of the suppression of the unwanted signals at the IF output. The isolation can vary over a large range, with most having an isolation of 20–30 dB [2].

7.2 Using RF Photonics for Downconversion

A photonic-based downconverter seeks to replace a traditional electronic downconverter by providing increased isolation and frequency range, while improving the SFDR. One of the methods for using photonics for downconversion is replacing the various electronic components with photonic ones. An example of a photonic downconverter appears in Figure 7.3. The gain and filter stages can be easily replicated in the optical domain. The mixer is the piece that requires more analysis. One of the most straightforward ways to realize the mixing function is by using two cascaded optical modulators placed in a series configuration [3]. The generated optical sidebands are shown in Figure 7.3. The incoming RF signal has already been upconverted into the optical domain by an optical modulator at the antenna. A second modulator will use the LO as the input to the RF port. The resulting optical sidebands are present at the output of the modulator. The optical filter will then select out the pair of sidebands that match the desired IF. Then, the optical gain can be used to increase the conversion efficiency of the downconverter. Finally, the chosen LO and RF sidebands will be converted at the photodetector and sent to the electronic ADC for digitization.

The photonic downconverter offers some unique advantages. The optical modulator at the antenna acts like an upconverter, moving the RF signals into the optical domain. The same thing happens at the modulator in the downconverter for the LO. Since the LO and the RF are in the optical domain, they can only be detected when they beat with the optical carrier. Proper use of the optical filter will suppress the optical carrier and only allow the LO and RF optical sidebands to be presented to the photodetector. Ideally, the output will only be the IF, which is the difference in frequency between the LO and the RF. The result makes the isolation of the downconverter essentially infinite, as the LO and RF frequencies do not appear at the output of the downconverter. The other advantage of the optical filter is the ability to suppress the image frequency. The LO can be set near one edge of the filter, and the image frequency of the RF signal to be downconverted will be outside the passband of the filter. This is also true for the image frequency on the other side of the optical carrier. Another advantage of the photonic downconverter is the wide bandwidth operation. The photonic modulator can be used with an LO frequency as high as 100 GHz. The use of photonic-based oscillators, as described in Chapter 4, can generate HF LOs with low phase noise. Therefore, the photonic downconverter can operate at frequencies much higher than most conventional electronic solutions.

The photonic downconverter does have some limitations. As discussed in Chapter 2, the RF gain of a photonic link is directly related to the DC

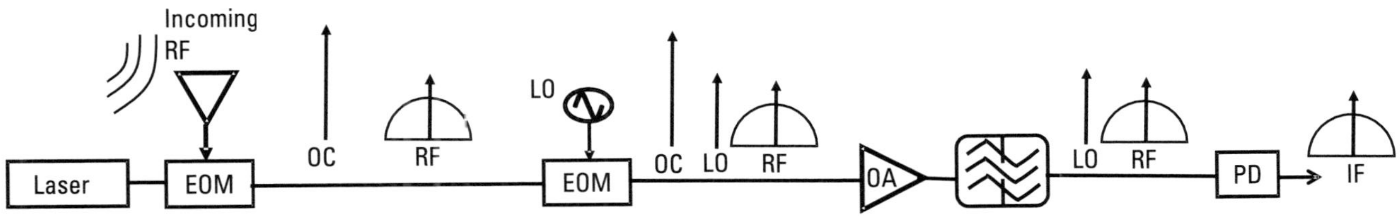

Figure 7.3 Block diagram of a serial cascaded modulator photonic RF downconverter (EOM: electro-optic modulator; OC: optical carrier; OA: optical amplifier; and PD: photodiode). (After: [1].)

photocurrent squared. The DC photocurrent is directly proportional to the optical power that is received by the photodetector. In the serial modulator downconverter configuration, the optical carrier and all other sidebands are filtered out, significantly reducing the total optical power at the photodetector. Therefore, the conversion efficiency of the downconverter will be lower. While using an optical amplifier can help improve the conversion efficiency, it will still be low. In addition the optical amplifier will add noise that will also cause the RF noise figure to be higher. Finally, the optical modulator has its own nonlinear transfer function, which can reduce the SFDR of the down converted signals.

7.3 Advancements in RF Photonic Downconverters

In order to address some of the deficiencies of the RF photonic downconverter, different architectures have been developed. In one example, the DC bias of the modulator used for the LO has been optimized to cancel the distortion generated from the modulator at the antenna [4]. The results show that either the NF or the SFDR can be improved but not both at the same time. In another configuration, the intensity modulators are replaced with phase modulators [5]. The advantage of this configuration is the lower optical insertion loss for each of the modulators as well as the lack of a bias control circuit to hold the modulator at the appropriate operating point. The lower optical losses result in a relative increase in the RF gain of 2.6 dB when compared to an intensity-modulated photonic link. The link when set for the optimal SFDR performance demonstrates an increase of 10.5 dB. This comes at a reduction of the conversion efficiency of 13.6 dB.

Beyond using modulators in series-cascaded architecture, a parallel modulator scheme can be employed for downconversion [6]. As seen in Figure 7.4, the laser output is split along two parallel paths. The upper path contains the modulator at the antenna, while the lower path contains the modulator that is connected to the LO. The two outputs of the modulators are then filtered to get the LO and RF sidebands. The two arms are then combined at a 50/50 coupler, and the two outputs are connected to a balanced pair of photodiodes. This architecture allows for cancelation of common mode noise from the laser, which can improve the RF NF. The separate paths also provide more optical power at the photodiodes, which allows for a higher conversion efficiency. In fact, a comparison of the parallel and series architectures for downconversion summarizes the improvements of the parallel architecture over the cascaded series configuration [7]. The parallel architecture allows for suppressed carrier operation of the modulators, allowing for improved

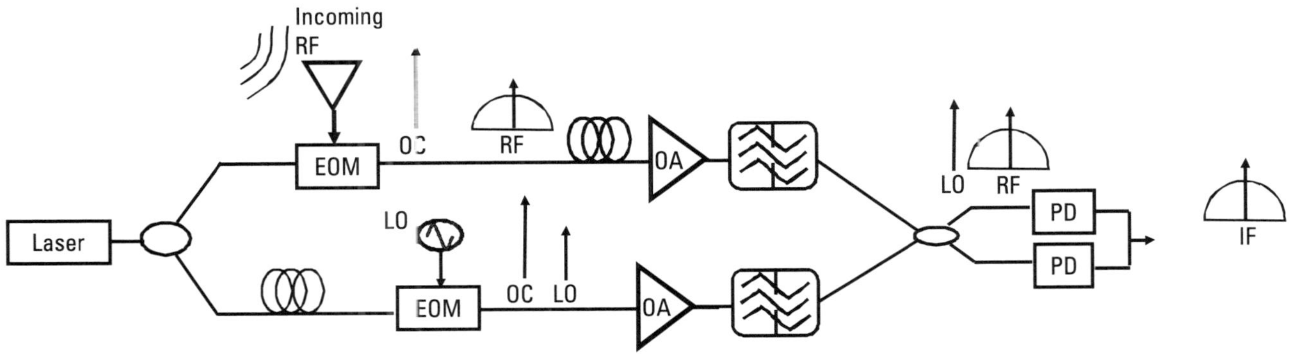

Figure 7.4 Block diagram of a parallel modulator photonic RF downconverter (EOM: electro-optic modulator; OC: optical carrier; OA: optical amplifier; and PD: photodiode). (After: [1].)

performance. The measured results show that the parallel architecture offers an improvement in the SFDR of 14 dB while offering a conversion efficiency 4 dB higher. The comparison does note that the drawbacks of using the parallel setup. The system looks like a large interferometer. It requires the two paths to be matched on optical scales. Any variation in the two paths, due to either temperature or vibration, will cause the system to become unstable and result in added noise. An active control to compensate for changes in the path lengths must be implemented. Thus, the added improvements come at an added complexity as well.

In order to overcome the deficiencies of the parallel architecture while maintaining the improved performance, the use of dual-parallel MZMs (DP-MZMs) has been proposed [8]. The DPMZM has been described in previous chapters and appears as an integrated form of the parallel-down converting setup. The DPMZM also has a phase control in one of the parallel arms of larger interferometer structure. The use of this structure has been shown to have a conversion efficiency of 7.8 dB, which is a 23.7-dB improvement over a series-cascaded downconverter operating at similar conditions. The DPMZM's susceptibility to environmental fluctuations is much lower than a fiber-based approach, because in the former, the paths are very short and confined to a single chip. The structure does have some trade-offs. First, the structure does not have balanced outputs to cancel the common mode noise. The demonstration does show a higher noise floor when compared to the serial cascade setup. Second, the rejection of the LO is not infinite. While it is 45 dB, it is not as high as in the other optical configurations. Finally, since the parallel structure is integrated into a single modulator, the RF and the LO have to be colocated.

In the demonstration architecture, the RF is received at the antenna while the LO is generated at the processing block. The LO can be remoted to the antenna site, but this leads to added complexity. To address the LO generation at the DPMZM, a demonstration utilizing an OEO has been made [9]. The DPMZM is incorporated into an OEO feedback cavity. The RF signal is still connected to an antenna, but the LO port is connected to the OEO feedback. A low phase noise LO can be generated right at the DPMZM, bypassing the need to remote an LO to the antenna. Unfortunately, the OEO cavity is not widely and continuously tunable, making it ideal for downconverting fixed-frequency RF signals but not able to downconvert other frequencies. Clearly, different setups exist for downconverting RF signals. The ideal one often depends on the application and the required performance. In most cases, the trade-off comes between the RF NF and SFDR. Optimizing one will often sacrifice the performance of the other.

7.4 RF Photonic Analog-to-Digital Conversion

While the RF photonic downconverter has great utility for the demonstration architecture it is still limited by the electronic ADC. For an ADC, there are often three important noise factors that limit the SNR performance of the digitizer. The first is the noise floor of the ADC, which is a combination of the thermal noise limit for a given bandwidth and the added noise of the ADC itself. The second is the aperture jitter, which is due to the noise in the sample and hold circuit of the ADC. Finally there is the clock jitter, which is due to the sampling clock. Previous work often assumed that the jitter was white noise and uniform. However, it has been shown that this does not hold true, as the phase-locked loop-based synthesizers used as the clock source shape the noise by the loop transfer function [10]. Because of this, the clock jitter is the dominant noise source for high sampling frequencies [11]. One of the reasons for this is that the aperture's jitter-dependent SNR can be improved by oversampling, whereas the SNR due to the sampling jitter cannot. With this limitation in mind, recent advancements in ADC technology have led to wider analog bandwidths being available. One of the methods for increasing the bandwidth of the ADCs involves taking lower bandwidth ADCs and parallelizing them in order to increase the overall bandwidth. The bandwidth of the composite ADC improves by N times the bandwidth of an individual ADC, where N is the number of ADCs used in the parallel structure. The use of the parallel ADC comes with some added complexities. The total SNR performance of the parallel ADC is shown in (7.1) [12].

$$ENOB_{total} = \frac{1}{6.02}\left[\frac{1}{\left(\left(2\pi f_{bw}\sigma_{tj}\right)^2 + \frac{\left(2\pi f_{bw}\sigma_{aj}\right)^2}{N} + \frac{1}{N \times SNR_{th}}\right)} - 1.78\right] \quad (7.1)$$

where f_{bw} is the 3-dB bandwidth of the individual ADCs, SNR_{th} is the thermal-limited SNR, σ_{tj} is the timing jitter of the sampling clock, and σ_{aj} is the aperture jitter. The SNR improves as a function of N, while the aperture jitter also improves as a function of N. However, the sampling jitter does not improve as the number of ADCs increases, as the same clock is used for all of the ADCs. Since the clock is the same, the timing jitter is correlated between the various ADCs and does not scale with *N*. As seen in Figure 7.5, the timing jitter will limit the performance of the parallel ADC structure as the input bandwidth increases. Changing the phase of the incoming signal to each of the parallel ADCs has been shown to improve the overall performance by only 3 dB or 0.5 ENOB [12]. Another improvement would be to make an independent clock for each of the ADCs in parallel. While this would allow for some decorrelation of the noise, it yields some other problems. For example, clock skew between the various clocks will create strong peaks at harmonic multiples of the sampling clock frequency [13]. This will limit the overall usable bandwidth of the composite ADC.

Note that the above analysis examines SNR improvements without considering the nonlinear dynamic range. By using a parallel structure, the ideal system would provide an improvement of *N* to the dynamic range as compared to a single ADC. This assumes that the nonlinear responses of the ADCs are completely decorrelated. Previous work has shown that this is not the case [14]. While predistorting the signal before each ADC and then undoing the response has been shown to improve 2x2 spurs, the third-order intermodulation shows no improvement. Parallel ADC structures show promise for improving the overall bandwidth and ENOB while making use of lower-sampling speed ADCs. Photonic techniques may be able to improve some of the limitations of the electronic solutions.

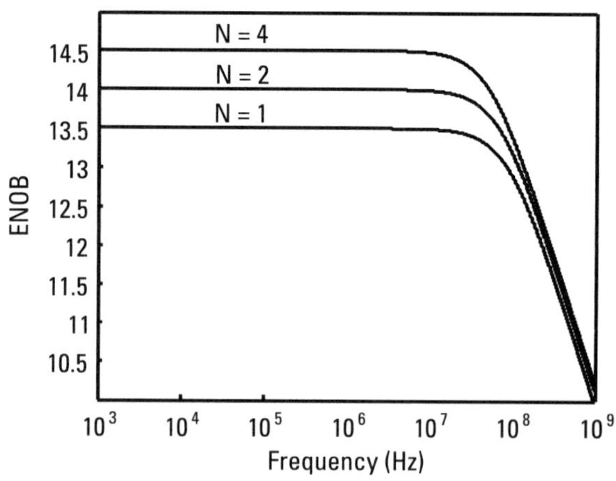

Figure 7.5 ENOB as a function of frequency for a single (N = 1) ADC, for two (N = 1), and four (N = 4) parallel-configured ADCs. The parallel ADCs improve the ENOB at low frequencies but become limited at higher frequencies due to the timing jitter of the sampling clock. (After: [1].)

RF photonics brings two important features for ADCs: wide bandwidth combined with low sampling jitter. In an in-depth review paper [15], four classes of photonic ADCs are identified: photonic-assisted, photonic-quantized, photonic-sampled, and photonic-sampled and -quantized. Each class is described with examples [16–19]. The main difference between each class is how RF photonics are utilized while covering the advantages and disadvantages of the various implementations. For this discussion, the focus will be placed on photonic sampled as well as photonic sampled and quantized. The photonic-sampled case uses techniques in the photonic domain to sample the incoming RF signal so that one or multiple electronic ADCs can digitize the signal. This class is one of the most mature and has the highest performance to date. It is also the most closely related to the parallel electronic structure. On the other hand, photonic sampling and quantization is the most ambitious as it seeks to bypass any electronic ADC and perform all functions before finally using photodetection to create the bits. It is also the least mature and offers the most potential growth.

7.5 RF Photonics Sampling Combined with Electronic Quantization

The use of photonics for the sampling function takes advantage of the wide bandwidth operation as well as the low timing jitter. An RF photonic sampling scheme is shown in Figure 7.6. The use of mode-locked lasers has been able to generate a high repetition rate with very low pulse-to-pulse jitter [17]. The optical pulses are then modulated by the RF signal to be sampled. The modulated optical pulses can then be time-demultiplexed and sent to multiple ADCs [20]. The configuration has the advantages of the electronic

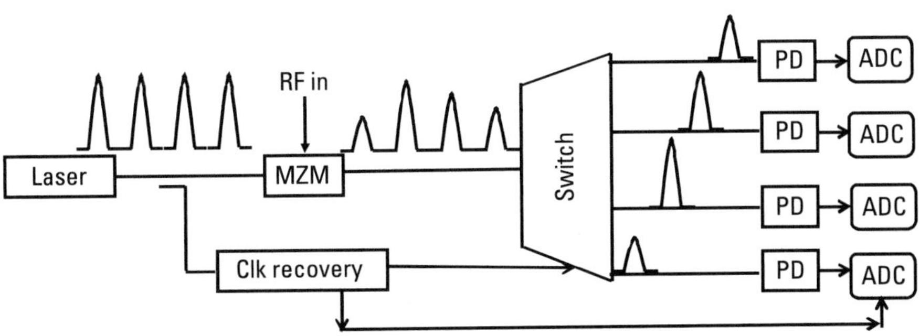

Figure 7.6 Block diagram of a photonic ADC using optical sampling and electronic quantization with time-division multiplexing using an optical switch. (After: [1].)

parallel ADC while improving the sampling clock jitter. However, the setup requires strict calibration to ensure that the time-demultiplexing is properly maintained. The use of optical switches for the demultiplexing also adds extra loss—and requires a time synchronization between the laser and the switch. All of these complexities can degrade the ENOB performance of the photonic ADC [15]. An improvement in the system uses multiple wavelength pulse sources, as seen in Figure 7.7. The use of a WDM reduces the losses and simplifies the timing synchronization [21].

In either case, the laser plays a significant role in how well the photonic ADC operates. Since the sampling function is done in the photonic domain, the laser pulse width, timing jitter, and amplitude variations will play directly into the overall performance. The ENOB as a function of pulse width and sampling frequency is given by [22]

$$ENOB_{pw} = \log_2\left(\frac{4}{\pi\left(f_{samp}\tau_{pw}\right)^2\sqrt{6}}\right) \tag{7.2}$$

where f_{samp} is the sampling frequency and τ_{pw} is the sampling optical pulse width. The timing jitter is given by the Walden relation [15, 23]

$$ENOB_{tj} = \log_2\left(\frac{1}{\pi f_{samp}\sigma_{tj}\sqrt{6}}\right) \tag{7.3}$$

where σ_{tj} is the timing jitter. Finally the amplitude variations have to be less than the quantization error for the least significant bit as given by [15]

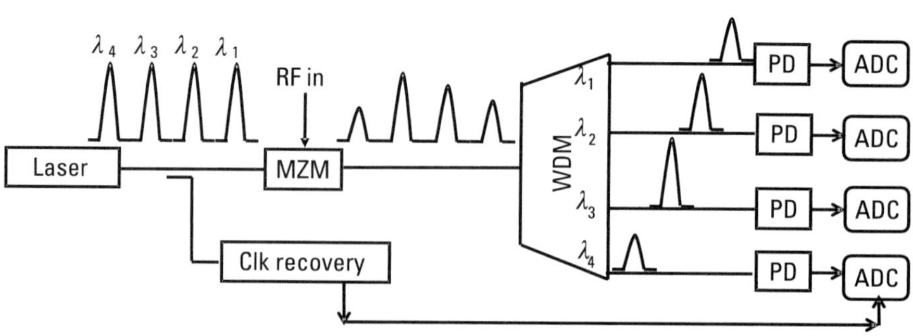

Figure 7.7 Block diagram of a photonic ADC using optical sampling and electronic quantization with wavelength-division multiplexing using an optical demux. (After: [1].)

$$ENOB_{amp} = \log_2\left(\frac{1}{\sigma_{amp}\sqrt{6}}\right) \tag{7.4}$$

where σ_{amp} is the amplitude ripple. All of these relations have to be considered to optimize the ENOB performance for a given sampling frequency. Finally, the laser clock signal has to be recovered so that it can be used to synchronize the clock of the electronic ADCs in order to realize the improved performance.

Another method is to use a time stretch method. In this case the optical source is one that can generate a chirped waveform as seen in Figure 7.8. Ideally the waveform has been modified such that each of the individual wavelengths in the pulse has been continuously spread in time [24]. The waveform can then be passed through an intensity modulator, which will impart the RF signal onto the amplitude of the waveform. Then a second stage will time-stretch the waveform. By proper configuration of the second stage, a wavelength demultiplexer can be used to create overlapping time blocks of the RF signal. The separated signals can then be quantized by a set of parallel electronic ADCs. The resulting digital data then has to be interleaved to recover the original RF signal.

The time-stretch photonic ADC has considerations similar to those of the optical pulse-sampled system. The optical paths have to be well matched in both length and loss. However, it has been noted that the overlap in signals across the ADCs can help simplify the calibration needed to coordinate the multiple paths [25]. In addition, the use of dispersion to both create the chirped waveform as well as to time-stretch the RF-modulated optical signal places limits on the bandwidth of the system [24]. As described in Chapter 3, the use of dual-sideband modulation places a limit on the frequency as dispersion will result in RF fading. While single-sideband techniques have been demonstrated to overcome these issues, they can limit the frequency bandwidth of the system.

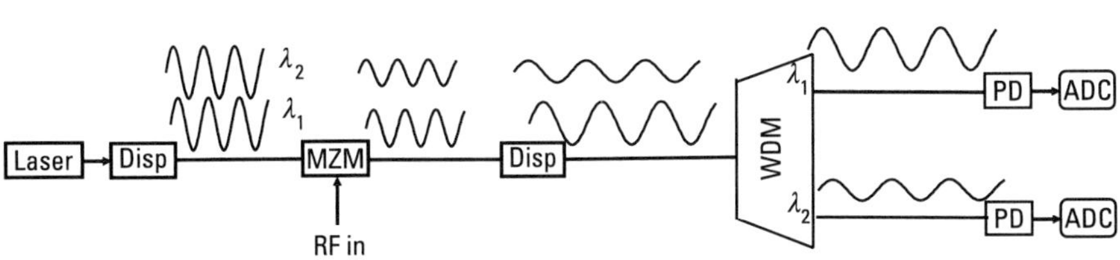

Figure 7.8 Block diagram of a time-stretch photonic ADC using optical sampling and electronic quantization with wavelength-division multiplexing using an optical demux. (After: [2].)

7.6 Photonics Sampling and Quantization

Photonic ADCs that include photonic quantization. The quantization that occurs in the previous discussion uses electronic solutions that are limited by the comparators that are used in the electronic ADCs. Instead, Taylor [19] proposed to use an array of modulators to perform the quantization function. As seen in Figure 7.9, each of the modulators is identical except for the length of the RF electrodes. By setting the electrodes of one of the modulators to half the length of the other, a two-bit quantizer can be realized, as seen in the inset of Figure 7.9. However, the system requires that not only the length of the electrodes be half, but that the half-wave voltage, or V_π, be half as well [15, 19]. So if the first modulator has a V_π of 6V, the second one must have a voltage of 3V. This requirement has limited the demonstration of this configuration to only 4 bits. Other demonstrations have been designed to improve the performance of this initial system. Instead of a parallel array, serially cascaded intensity [26] and phase modulators [27] have been demonstrated. Also, a parallel array of modulators with equal-length electrodes has been proposed [28]. The quantization is achieved by operating each modulator at a different DC bias point in order to create an intensity-dependent response. Polarization has also been used to create a photonic quantizer [29].

Beyond the used of multiple modulators, other methods for photonic quantization have been explored. Nonlinear optical responses can be exploited to perform a thresholding effect on the RF signal for quantization. The thresholder will take the intensity of the optical signal and compare it to a set quantization level. If the intensity is greater than the level, it will register as a 1. Otherwise it will register as a 0. While many examples of nonlinear optical responses can be used, one method uses a saturable absorber to perform the thresholding operation [30]. As seen in Figure 7.10, the saturable absorber will block the transmission of light when the input optical power is below a certain level. However, when the optical power increases, the saturable

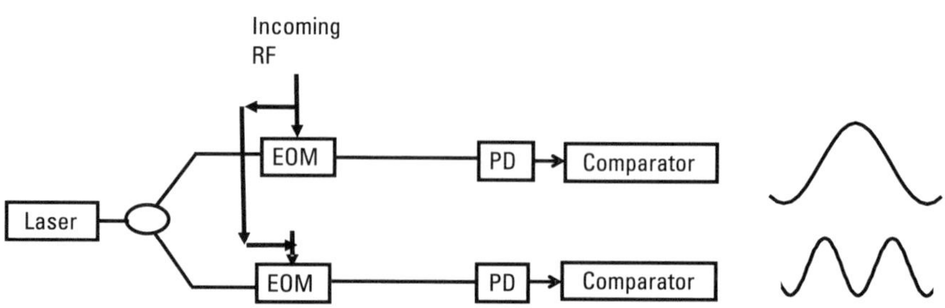

Figure 7.9 Block diagram of a photonic ADC using optical quantization with two bits of resolution. (After: [1].)

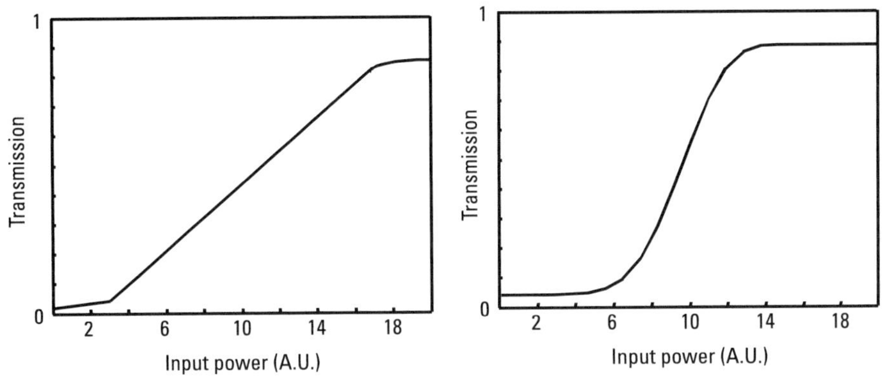

Figure 7.10 Transmission response as a function of input power for a saturable absorber (left) and a saturable absorber in an optical cavity (right). (After: [1].)

absorber will begin to allow the optical power to pass with lower loss. Unfortunately, the slope of the response is too shallow for use as a thresholder. In order to improve the slope, the saturable absorber can be placed in an optical cavity to increase the slope. Ideally, the cavity will allow the incoming light to pass through the cavity multiple times, causing the response to become sharper, as seen in Figure 7.10. The increase in the response comes with an increased overall transmission loss, limiting the optical power for detection that will be used to derive the bits for the ADC. Other nonlinear optical responses have been investigated, including four-wave mixing [31] and nonlinear Sagnac interferometers [32].

7.7 Arbitrary-Transmit Waveform Generation Using RF Photonics

The demonstration architecture incorporates a circulator in order to allow the antenna to be used for both transmission and reception. To generate simple RF sinusoidal signals, the oscillators described in Chapter 4 can be used as very HF signal sources. They can either have very low phase noise or can be tuned over very wide frequency ranges. In fact a signal source has been demonstrated that can generate signals up to 100 GHz [33]. On the other hand, some applications may require more complex types of waveforms. Examples include communications, medical imaging, and test and measurement systems, along with predistortion of signals to compensate for the nonuniform impulse response of the antenna itself [34]. All of these can be accomplished with photonic-based arbitrary waveform generation. This section reviews a few different methods for arbitrary waveform generation.

A review of arbitrary waveform generation using photonics highlights some of the earlier works that utilized a spatial light modulator (SLM) in a free-space configuration [35]. As seen in Figure 7.11, an optical source that has a large optical bandwidth is input to the SLM. The SLM then acts as a spectral filter or shaper. It modifies the amplitude and phase of the different parts of the spectrum. The light is then passed down a medium with a linear dispersion profile. The dispersion is used to convert the wavelength of the optical spectrum into the time domain. Note that the use of dispersion is similar to the time-stretch process described in the photonic ADC section, Section 7.5. Finally, the optical signal is input to a photodetector, and the resulting time-varying signal can be sent to the antenna. This system is a combination of wavelength-to-space mapping, followed by wavelength-to-time mapping techniques to generate the required waveform. The SLM can be implemented using many different techniques [36]. The use of a liquid-crystal array has been used most frequently, due to its abilities to be quickly reconfigured and to control both the phase and amplitude of the light at each input to the array.

This system has been shown to generate very wideband arbitrary waveforms. Unfortunately, the use of a free-space system can be difficult to stabilize. An advancement of the previous demonstration is to replace the gratings with a fiber-based arrayed-waveguide grating (AWG) to spread the optical spectrum before the SLM [37]. The AWG can have the various channels of the spectrum travel different lengths along with individual attenuators, allowing for control of the phase and the amplitude as in the other configuration.

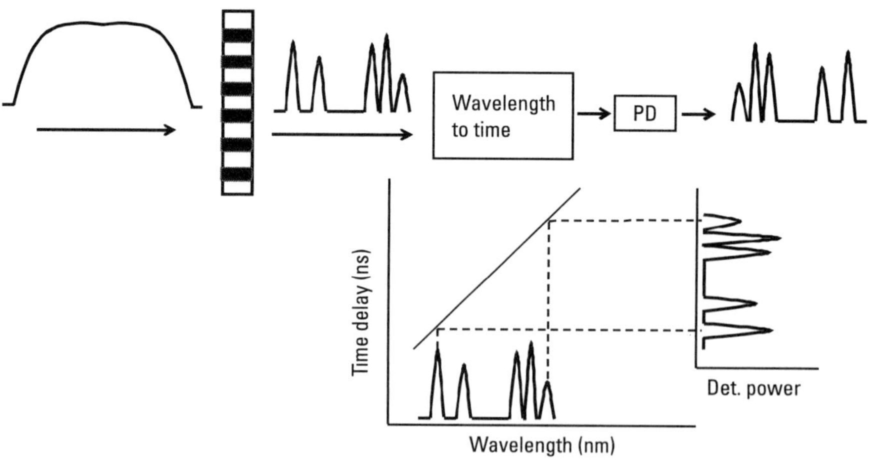

Figure 7.11 Photonic arbitrary waveform generation using a broadband source with a spectral shaper followed by wavelength-to-time mapping using dispersion. (After: [1].)

The advantages of this configuration are greater stability with a smaller footprint. On the other hand, the resolution of a typical AWG is around 12 GHz, while a free-space grating can have a resolution of a few gigahertz. This resolution will limit the performance of the system in terms of the waveform that is generated.

Beyond the use of an AWG or an SLM for the spectral shaper, other methods have been suggested. In one case, a linearly chirped Fiber-Bragg grating (FBG) can be used to perform the proper modification of both the phase and amplitude [38]. In fact, the linearly chirped grating can be used for both the spectral shaping and the wavelength-to-time mapping [39]. Unfortunately, in both of these cases, the grating cannot be easily changed, thus making the system unable to be dynamically configured for new waveform generation.

Another improvement on the previous demonstrations involves the optical source used. The previous methods depend on either a supercontinuum source or a mode-locked laser that generates very short optical pulses. In either case, much of the power will be lost as it is removed by the SLM or AWG. In order to generate a strong RF signal after detection, the optical power will have to be quite high before entering the system. In addition, the optical spectrum must ideally be flat in amplitude and have the phase of the optical wavelengths be a fixed relationship. Another source that has been suggested for use in waveform generation is an optical frequency comb source [40]. The optical comb source provides some advantages. The comb lines can be set to match the channel spacing of the grating or the AWG in order to perform phase and amplitude control on each of the comb teeth individually. This will allow for no additional optical power to be lost. The resulting signal will then be used at the photodetector to generate an RF signal with the desired waveform.

Other methods for generating arbitrary waveforms exist beyond that of the spectral filtering followed by wavelength-to-time transform. In some cases, an optical pulse can be time-multiplexed to generate time-delayed versions of itself. When the pulses are detected at a photodetector and the electrical signal is passed through a low-pass filter, the generated RF waveform will be a smooth, continuously time-varying RF signal [41]. Another method uses an optical modulator combined with higher-order dispersion effects to generate higher frequency-chirped RF waveforms from a low-frequency RF signal at the modulator input [42]. Finally photonic-based microwave filters can be designed as matched filters in order to generate specialty waveforms for communication applications [43].

7.8 Conclusions

This chapter expands the processing-block demonstration architecture to include three different systems: a photonics-based RF downconverter, an ADC, and an arbitrary-waveform generator. Once a signal has been identified for further investigation, the RF photonic downconverter can be tuned to the signal frequency and mix the signal down for digitization by a low sampling frequency electronic ADC. As the signals received by the demonstration architecture can have very high center frequencies, a wideband photonic RF downconverter will be well suited for this application. Multiple methods for performing downconversion of RF signals are presented, each with their own advantages and disadvantages. The choice of which one to use depends on the overall application and need.

The photonic-based ADC seeks to overcome the limitations of the traditional electronic-based ADC. As discussed previously, one of the main limiters for wideband ADCs is the required sampling clock timing jitter. Photonics has long been shown to provide sampling sources with timing jitter on the order of tens of femtoseconds. When combined with arrays of low-bandwidth quantizers, demonstrations have shown sampling of wideband signals with a high effective number of bits. These hybrid photonic-electronic ADCs have become mature in their performance. On the other hand, fully photonic-based sampling and quantization remains a field with room to grow. Photonic quantization has been shown to be achievable, but with only a limited number of bits. Work continues to explore advancements in this area and may lead to future developments that make photonic ADCs with much higher performance.

The photonic-based arbitrary-waveform generator can be used to create unique waveforms for transmission from the demonstration architecture. Multiple implementations have been shown for generating waveforms with very wide bandwidth. Many of them use a spectral filter or shaper in order to operate on individual wavelengths of a broad spectrum optical source. By separating the individual components of the broader optical spectrum, individually modifying the phase and amplitude of those components, and then recombining the components, a new spectrum can be generated. Converting the spectrum to a time-varying signal and then photodetecting the optical signal generates the required RF signal. While some methods have used fixed spectral filters, others have incorporated reconfigurable amplitude and phase control, allowing the RF signal to be dynamically modified. All of these subsystems allow the demonstration architecture to measure, record, and transmit wideband signals.

References

[1] Devgan, P., "Signal Processing Utilizing Radio Frequency Photonics," *AFRL Report,* AFRL-RY-WP-TR-2017-0172, 2017.

[2] Marki, F., and C. Marki, "Mixer Basics Primer," Marki Microwave Inc., Application Note, 2010.

[3] Gopalakrishnan, G., W. Burns, and C. Bulmer, "Microwave-Optical Mixing in LiN-bO_3 Modulators," *Transactions on Microwave Theory and Techniques,* Vol. 41, No. 12, 1993, pp. 2383–2391.

[4] Karim, A., and J. Davenport, "High Dynamic Range Microwave Photonic Links for RF Signal Transport and RF-IF Conversion," *Journal of Lightwave Technology,* Vol. 26, No. 15, 2008, pp. 2718–2724.

[5] Pagan, V., B. Haas, and T. Murphy, "Linearized Electrooptic Microwave Downconverter Using Phase Modulation and Optical Filtering," *Optics Express,* Vol. 19, No. 2, 2011, pp. 883–895.

[6] Burns, W., G. Gopalakrishnan, and R. Moeller, "Multioctave Operation of Low-Biased Modulators by Balanced Detection," *Photonics Technology Letters,* Vol. 8, No. 1, 1996, pp. 130–132.

[7] Gallo, J. and J. Godshall, "Comparison of Series and Parallel Optical Modulators for Microwave Down-Conversion," *Photonics Technology Letters,* Vol. 10, No. 11, 1998, pp. 1623–1625.

[8] Chan, E., and R. Minasian, "Microwave Photonic Downconverter with High Conversion Efficiency," *Journal of Lightwave Technology,* Vol. 30, No. 23, 2012, pp. 3580–3585.

[9] Tang, Z., F. Zhang and S. Pan, "Photonic Microwave Downconverter Based on an Optoelectronic Oscillator Using a Single Dual-Drive Mach-Zehnder Modulator," *Optics Express,* Vol. 22, No. 1, 2014, pp. 305–310.

[10] Da Dalt, N., et al., "On the Jitter Requirements of the Sampling Clock for Analog-to-Digital Converters," *Transactions on Circuits and Systems I: Fundamental Theory and Applications,* Vol. 49, No. 9, 2002, pp.1354–1360.

[11] Löhning, M., and G. Fettweis, "The E ffects of Aperture Jitter and Clock Jitter in Wideband ADCs," *Computer Standards & Interfaces,* Vol. 29, No. 1, 2007, pp. 11–18.

[12] Lauritzen, K., S. Talisa, and M. Peckerar, "Impact of Decorrelation Techniques on Sampling Noise in Radio-Frequency Applications," *Transactions on Instrumentation and Measurement,* Vol. 59, No. 9, 2010, pp. 2272–2279.

[13] Kobayashi, H., et al., "Aperture Jitter Effects in Wideband Sampling Systems," *Proc. of Instrumentation and Measurement Technology Conference,* 1999.

[14] Howard, L., and D. Rabideau, "Correlation of Nonlinear Distortion in Digital Phased Arrays: Measurement and Mitigation," *Microwave Symposium Digest,* 2002.

[15] Valley, G., "Photonic Analog-to-Digital Converters," *Optics Express,* Vol. 15, No. 5, 2007, pp. 1955–1982.

[16] Jacobs, E., et al., "Optically Clocked Track-And-Hold for High-Speed High-Resolution Analog-to-Digital Conversion," *International Topical Meeting on Microwave Photonics,* 2004.

[17] DePriest, C., et al., "High-Quality Photonic Sampling Streams from A Semiconductor Diode Ring Laser," *Journal of Quantum Electronics,* Vol. 38, No. 4, 2002, pp. 380–389.

[18] Zmuda, H., "Analog-to-Digital Conversion Using High-Speed Photonic Processing," *SPIE International Symposium on Optical Science and Technology,* 2001, pp. 84–95.

[19] Taylor, H., "An Optical Analog-to-Digital Converter—Design and Analysis," *Journal of Quantum Electronics,* Vol. 15, No. 4, 1979, pp. 210–216.

[20] Bell, J., et al., "A/D Conversion of Microwave Signals Using a Hybrid Optical/Electronic Technique," *SPIE Optical Technology for Microwave Applications,* Vol. 1476, 1991, pp. 326–329.

[21] Juodawlkis, P., et al., "Optically Sampled Analog-to-Digital Converters," *Transactions on Microwave Theory and Techniques,* Vol. 49, No. 10, 2001, pp. 1840–1853.

[22] Valley, G., J. Hurrell, and G. Sefler, "Photonic Analog-to-Digital Converters: Fundamental and Practical Limits," *Proc. of SPIE,* Vol. 5618, 2004, pp. 96–106.

[23] Walden, R., "Analog-to-Digital Converter Survey and Analysis," *Journal on Selected Areas in Communications,* Vol. 17, No. 4, 1999, pp. 539–550.

[24] Han, Y. and B. Jalali, "Photonic Time-Stretched Analog-to-Digital Converter: Fundamental Concepts and Practical Considerations," *Journal of Lightwave Technology,* Vol. 21, No. 12, 2003, pp. 3085–3103.

[25] Han, Y., and B. Jalali, "Continuous-Time Time-Stretched Analog-To-Digital Converter Array Implemented Using Virtual Time Gating," *Transactions on Circuits and Systems I: Regular Papers,* Vol. 52, No. 8, 2005, pp. 1502–1507.

[26] Jalali, B., and X. Yie, "Optical Folding-Flash Analog-to-Digital Converter with Analog Encoding," *Optics Letters,* Vol. 20, No. 18, 1995, pp. 1901–1903.

[27] Currie, M., T. Clark, and P. Matthews, "Photonic Analog-to-Digital Conversion by Distributed Phase Modulation," *Photonics Technology Letters,* Vol. 12, No. 12, 2000, pp. 1689–1691.

[28] Chi, H., and J. Yao, "A Photonic Analog to Digital Conversion Scheme Using Mach Zehnder Modulators with Identical Half Wave Voltages," *Optics Express,* Vol. 16, No. 2, 2008, pp. 567–572.

[29] Li, W., et al., "All Optical Analog to Digital Conversion Based on Polarization-Differential Interference and Phase Modulation," *Photonics Technology Letters,* Vol. 19, No. 8, 2007, pp. 625–627.

[30] Hayduk, M., R. Bussjager, and M. Getbehead, "Photonic Analog to Digital Conversion Techniques Using Semiconductor Saturable Absorbers," *Proc. of SPIE,* Vol. 4042, 2000, pp. 54–60.

[31] Oda S., A. Maruta, and K. Kitayama, "All-Optical Quantization Scheme Based on Fiber Nonlinearity," *Photonics Technology Letters,* Vol. 16, No. 2, 2004, pp. 587–589.

[32] Kitayama, K., et al., "Photonic Analog-to-Digital Conversion," *Digest of the LEOS Summer Topical Meetings,* 2005.

[33] Schneider, G., et al., "Radiofrequency Signal-Generation System with Over Seven Octaves of Continuous Tuning," *Nature Photonics,* Vol. 7, No. 2, 2013, pp. 118–122.

[34] McKinney, J. and A. Weiner, "Compensation of the Effects of Antenna Dispersion on UWB Waveforms Via Optical Pulse Shaping Techniques," *Transactions on Microwave Theory and Techniques,* Vol. 54, No. 4, 2006, pp. 1681–1685.

[35] Yao, J., "Photonic Generation of Microwave Arbitrary Waveforms," *Optics Communications,* Vol. 284, No. 15, 2011, pp. 3723–3736.

[36] Goswami, D., "Optical Pulse Shaping Approaches to Coherent Control," *Physics Reports,* Vol. 374, No. 6, 2003, pp. 385–481.

[37] Fontaine, N., et al., "Compact 10 GHz Loopback Arrayed Waveguide Grating for High Fidelity Optical Arbitrary Waveform Generation," *Optics Letters,* Vol. 33, No. 15, 2008, pp. 1714–1716.

[38] Wang, C. and J. Yao, "Photonic Generation of Chirped Microwave Pulses Using Superimposed Chirped Fiber Bragg Gratings," *Photonics Technology Letters,* Vol. 20, No. 11, 2008, pp. 882–884.

[39] Wang, C. and J. Yao, "Simultaneous Optical Spectral Shaping and Wavelength-to-Time Mapping for Photonic Microwave Arbitrary Waveform Generation," *Photonics Technology Letters,* Vol. 21, No. 12, 2009, pp. 793–795.

[40] Cundiff, S., and A. Weiner, "Optical Arbitrary Waveform Generation," *Nature Photonics,* Vol. 4, No. 11, 2010, pp. 760–766.

[41] Leaird, D. and A. Weiner, "Femtosecond Direct Space-to-Time Pulse Shaping in an Integrated-Optic Configuration," *Optics Letters,* Vol. 29, No. 13, 2004, pp. 1551–1553.

[42] Azana, J., et al., "Reconfigurable Generation of High-Repetition-Rate Optical Pulse Sequences Based on Time-Domain Phase-Only Filtering," *Optics Letters,* Vol. 30, No. 23, 2005, pp. 3228–3230.

[43] Bolea, M., et al., "Optical UWB Pulse Generator Using an N Tap Microwave Photonic Filter and Phase Inversion Adaptable to Different Pulse Modulation Formats," *Optics Express,* Vol. 17, No. 7, 2009, pp. 5023–5032.

8

Advancements in Integrated RF Photonics

The RF photonic subsystems that have been covered previously for use in a demonstration architecture have all been-based on discrete components that are connected by fiber-optic cables. As discussed in previous chapters, the performance of these components directly relates to the expected performance of the subsystems, from the relative intensity noise and optical power of the laser, to the electro-optic efficiency as a function of the RF frequency of the modulator and finally to the optical power handling and RF bandwidth of the photodetector. These and other discrete photonic components have been well developed by both commercial and private industries. In fact, advancements in these devices over the last 20 years has allowed for the previously described subsystems to meet the needs of various RF applications. Unfortunately, the components still suffer from drawbacks. One of the most important is the need for fiber-optic connections between the components. Most components come from commercial vendors with fiber-optic cable connections of at least one meter in length, and often as long as 2 or 3m. Figure 8.1 shows a photonic component with a 2-m length of fiber attached. For a simple link consisting of a laser, modulator, and photodetector, there can be as much as 12m of optical fibers connecting all the components. The length can be advantageous when the laser has to be physically placed away from the modulator, or the optical output of the modulator has to be routed to the front panel of a large box before it can be connected to a fiber-optic cable. However, the push has been to smaller and more integrated subsystems. Thus, the extra length of fiber consumes valuable space and often leads to complex routing inside the

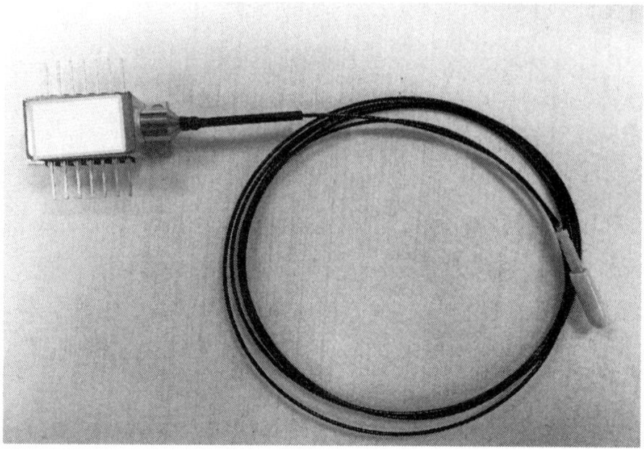

Figure 8.1 Photonic component with a fiber pigtail of 2m. (After: [1].)

subsystem box in order to minimize the space used. Another issue with the fiber-optic connections comes with their sensitivity to the local environment. When the laser output is split into two or more parallel paths, any thermal or vibration changes will be imparted onto the optical fiber. Often these changes cause the fiber to either stretch or contract, changing the path length that the light has to travel. If these changes are not matched or canceled out, they will impart fluctuations on the RF signal that is detected at the output of the subsystem, oftentimes degrading the fidelity of the signal.

A solution to these issues exists in integration. Being able to realize these components in an integrated photonic circuit can provide much smaller sizes while also making the subsystem impervious to environmental fluctuations. Integration of photonics supports both digital and analog applications. Integration does come with challenges. While discrete photonic components are mature, their integrated counterparts are still being improved. This chapter reviews integrated photonic components for RF photonics, along with subsystems that can benefit from the use of these circuits.

8.1 Integrated Photonic Fundamentals

When discussing integrated photonics, some basic definitions have to be set. Photonic integrated circuit (PIC) is a term that is often used in this area. For the purpose of this discussion, PICs are defined as photonic components that are monolithically integrated onto the same substrate. An example would be a laser integrated with an optical amplifier on a III-V semiconductor material platform. This provides a single component that can be used in a larger system. However, these PICs often require external electronic circuitry to power and control these components. In the above example, the laser and amplifier PIC are packaged and used as discrete, fiber-coupled components. The laser current source, along with the temperature controller, is separate from the PIC. On the other hand, integrated photonic circuits (IPCs) cover the area where multiple photonic and HF electronic components are integrated into a single circuit. For example, an IPC would only have an external power supply. The rest of this chapter focuses on IPCs and their use for RF photonic applications.

One of the ongoing debates in IPC is the choice of material upon which to fabricate the components. III-V-based semiconductor materials, such as indium phosphide, have been shown to support the photonic components necessary for making IPCs. The laser, modulator, and photodiode can all be fabricated on a single chip. In addition, HF modulators and photodiodes have been demonstrated in semiconductor materials [2, 3]. However the use of III-V materials comes with some caveats. The optical loss for III-V waveguides is larger than it is for silicon [4]. Also semiconductor modulators have been shown to have strong temperature-dependent performance [5]. Silicon-based integrated photonics (or silicon photonics for short) have been put forward as an alternative to semiconductor-based implementations. Being CMOS-compatible leverages the large foundry infrastructure that already exists for electronic integrated circuits. Along with this advantage, the optical propagation loss in silicon photonics has very low losses at 1,550 nm when compared to other material substrates. Germanium-based photodiodes with HF bandwidths have also been successfully fabricated on silicon [6]. Unfortunately, the use of silicon photonics comes with its own set of drawbacks. Unlike semiconductor-based IPCs, lasers at 1,550 nm are not easily achievable in silicon. Raman-based laser sources in silicon have been demonstrated, but require another optical sources to pump the laser [7]. The silicon photonic modulators that have been demonstrated depend on changing the free-charge carrier density in a PN diode configuration [8, 9]. This often leads to limits on the linearity performance of the modulator as well as the upper frequency operation.

Due to its low loss, silicon photonics appears best suited for use in IPCs that are mostly passive. These IPCs do not require a laser or optical amplification. Examples of these types of circuits are optical filters, switches, attenuators, wavelength multiplexers and demultiplexers, and optoelectronic feedback. Semiconductor-based IPCs provide more utility to active applications. A few examples include RF links, waveform generation, and RF oscillators based on optical injection locking. Unfortunately, no one material system seems to meet the needs of all applications.

The ideal solution would use heterogeneous integration in order to achieve the best performance. A heterogeneously integrated photonic circuit would be able to combine the HF operation of CMOS-based electronic circuits with the optical components afforded by semiconductor technology. The silicon waveguides and filters would also provide very low loss.

8.2 IPCs

Since RF photonic subsystems have been built on the low loss of optical fibers, silicon waveguides are preferred for these applications. Therefore, the use of IPCs will focus on technologies that can be used with these low-loss waveguides. The first component will be the laser. A germanium-based laser on silicon has been demonstrated with optical power on the order of a milliwatt [10]. For RF applications, the laser needs to deliver high optical power with low noise. Fortunately, the preferred discrete optical source is a semiconductor-based DFB laser. Work has focused on integrating the semiconductor laser with silicon waveguides. Direct bonding of the laser die to the silicon wafer has been demonstrated [11], with other methods focusing on using other types of adhesion [12, 13]. The bonding method has to take into account the different thermal expansion properties of the materials, which can lead to cracking or separation. In addition, the optical coupling between the laser and the silicon waveguide has to be properly designed in order to make the losses as small as possible.

The next component is the optical modulator. For analog applications, a high electro-optic conversion efficiency along with a low optical insertion loss will be desired. While a semiconductor-based modulator can be bonded with the silicon waveguides in a similar fashion as the laser, the lithium niobate ($LiNbO_3$) MZM has been the preferred modulator for RF photonics. Thin film $LiNbO_3$ (TFLN) has been developed in order to integrate it with silicon [14].The TFLN can be bonded to the silicon using an intermediate layer such as benzocyclobutene (BCB) that has low losses and will mitigate the thermal expansion difference between the two materials [15]. TFLN on silicon has been shown to have an electro-optic conversion efficiency and insertion loss close to bulk lithium niobate used in discrete modulators [16].

The optical amplifier is the next integrated component for discussion. Like the laser, the optical gain, along with a minimum amount of noise, is important. Again, semiconductor optical amplifiers can be bonded to the silicon waveguides. Still, the semiconductor-based amplifiers have two drawbacks when used for analog applications. The noise penalty for the semiconductor amplifier is around 8–26 dB, which is higher than the EDFA that is used in most analog applications. In addition, the semiconductor optical amplifier gain has a fast response time, leading to the gain being unintentionally modulated by the RF signal. This additional modulation can degrade the linear performance of the overall system. While an EDFA cannot be integrated, work has been done on doping the silicon waveguides with erbium [17]. The erbium-based amplifier will require a pump laser while the semiconductor amplifier can be electrically driven. The choice of amplifier will often be determined by the needs of the system.

For analog subsystems, integrated optical filters with high Q factors along with low insertion loss are needed, along with the ability to be tuned over a wide frequency range. Just as filters used in the discrete systems were based on FBGs, similar types of filters can be realized in the integrated photonic domain. Bragg gratings can be used in a FP configuration to provide an integrated filter, as seen in Figure 8.2. However, the optical insertion loss can be quite large for these devices. Another alternative is the ring resonator structure as seen in Figure 8.3. The ring resonator can act as both a bandpass and notch filter, much like the FBG [18, 19]. When light of a certain wavelength is coupled to the ring structure through the lower waveguide, it will simply pass through if it does not meet the resonant condition of the ring. The resonant wavelength is directly related to an integer multiple of the circumference of the ring. If the wavelength of the light does match the resonant condition, it will couple into the ring. The light will then couple out the upper waveguide. The upper waveguide will act like the bandpass while the lower waveguide will act like the notch filter. Ring resonators lend themselves to using multiple cascaded copies in order to improve the overall Q factor [20]. The resonant wavelength can also be tuned, similar to a FBG. The two methods are either heating [21] or straining by using a liquid-crystal material as the cladding for the ring resonator structure [22]. These types of filters can meet the various requirements of RF photonic subsystems.

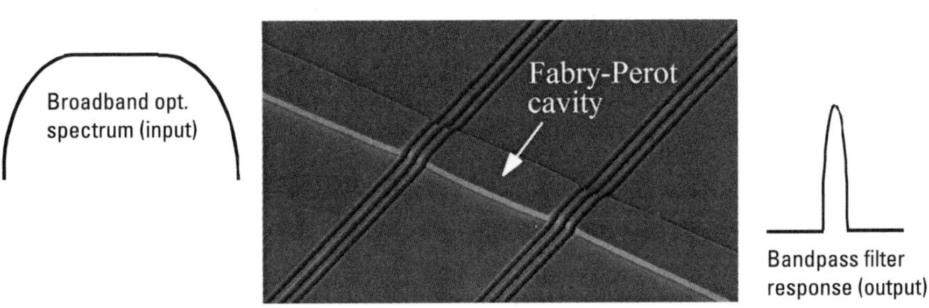

Figure 8.2 Integrated photonic FP filter. (After: [1].)

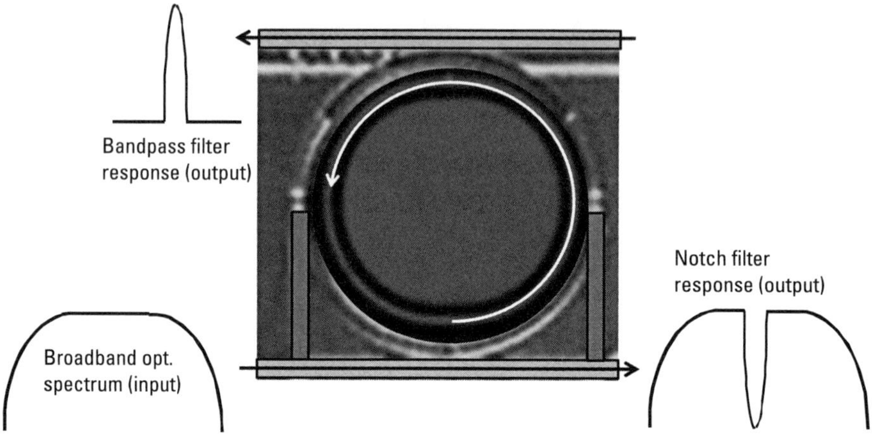

Figure 8.3 Integrated photonic ring resonator filter. (After: [1].)

Finally, the photodetector is the last component that needs to be realized for integrated RF photonic applications. The photodetector has to be able to handle high optical powers while maintaining a linear response over a wide frequency operating range. These types of photodetectors have been made in discrete forms and their performance continues to improve [23, 24]. One of the relevant improvements has occurred in how the light enters the device. Previous demonstrations have relied upon a surface normal interface to allow the light to reach the absorption region of the photodiode structure. As an improvement upon this device, photodetectors have started to use a waveguide interface [25]. In this case, the photodetector can be placed right above the waveguide, allowing the light to evanescently couple into the absorption region. The pursuit of these types of photodetectors can allow for higher frequency response and make it easier to integrate with silicon waveguides. In fact, high-power, HF photodetectors have been designed to integrate with silicon waveguides [26].

All of these heterogeneously integrated components should be able to provide the required performance for RF photonic subsystems. The main obstacle right now is the actual fabrication of these circuits in a single foundry setting. The integration will most likely require some form of 3-D structure, where the various components can be made separately and then stacked and bonded together. The key piece will then be the interconnections that link the different layers, allowing both the electrical and optical signals to move to the appropriate layer. The loss of these interconnects will be very important, as the analog performance strongly depends on the optical power that can be delivered to the photodetector. Work continues to improve performance in this area [27, 28].

8.3 Applications of IPCs to RF Photonics

Previous chapters have demonstrated RF photonic applications using discrete components. IPCs can be used in place of these systems. Starting with the full demonstration architecture shown in Figure 6.1, the various subsystems will be analyzed to see how IPCs can be used to replace the discrete component subsystems.

The RF circulator in the demonstration architecture is based on a MZM that is configured to isolate the transmitted signal at the antenna from the received one. The single-sideband modulation in the MZM requires the use of two 90° hybrids. The discrete hybrids can be quite large in size and are limited in the range of frequencies over which they operate. In an IPC, the hybrids can be made on the same chip as the MZM and optical filter. The resulting chip will be much smaller than its discrete version and can potentially be designed to have much wider frequency operation. In addition, the electronic bias control circuit needed to hold the MZM at the proper DC bias point can be included in the chip. Finally the laser can be also integrated onto the same chip. The RF circulator IPC could then be packaged in the same dimensions as a traditional electronic circulator, with the only difference being that the output port will be a fiber connection rather than an RF connection.

The next subsystem is the signal identification block. The spectrum analyzer examples from Chapter 7 can be realized in an IPC. The mixture of photonic and electronics would be a good candidate for this type of implementation. However, the same drawback of scanning the filter will still limit the utility of the circuit. Using the rare Earth-doped crystal removes the scanning drawback. Unfortunately, the crystal requires an external cooling system, potentially removing the advantages of miniaturization. Other methods of separating the different optical sidebands and using an array of photodiodes would seem better suited for IPCs. Just like the discrete version, the drawback is the separation that is achievable in the integrated filter.

The other frequency-identification demonstrations depend on passing the signal through either a dispersive element or a filter. Optical filters in IPCs can be accomplished either using Bragg gratings or ring resonators. The dispersive elements can be made in the waveguides as well. One method is to simply modify the waveguide design in order to induce a strong dispersive effect. Along the same lines, the waveguide can be doped with materials that will also increase the dispersion. Unfortunately, both ways may also increase the optical loss. The other method for signal identification is the MM-OEO. Unlike most OEOs, the MM-OEO tries to make the feedback loop as short as possible. This type of system seems well suited to an IPC, as it requires both photonic and electronic components that need to be placed as close together as possible. Along with the capability to amplify the signal, it is a strong candidate for an IPC implementation.

The next subsystem is the signal-separation block. The integrated optical filters can provide similar performance to their discrete counterparts. Still, the finite and infinite impulse response filters may be implemented in an IPC. The wavelength multiplexer and demultiplexer can be realized in silicon waveguides with very low loss [29]. The short delays can be implemented as well along with the amplitude control. The electronic circuit that will control each of the amplitude weights can also be integrated on the same chip. In fact other demonstrations have already been shown in integrated forms of the finite-impulse response filter [30].

The next block to be considered for implementation in an IPC is the oscillator. The MM-OEO has already been considered a good candidate for realization in an IPC. The standard OEO, however, requires a long delay line in order to achieve the low phase noise performance. One solution would be to integrate all parts of the OEO, but allow an output and input to the circuit that would allow for a long fiber delay to be added. Unfortunately, coupling the light out and back into the circuit can lead to extra loss, degrading the overall performance of the OEO. One way to avoid this loss is to create a cavity on the chip itself that can mimic the long fiber-delay line. An OEO with a whispering gallery mode optical resonator has been demonstrated that can produce high-quality, low phase noise RF signals without the need for a long fiber-delay line [31]. As for oscillators based on optical injection locking of two lasers, these setups require an integrated optical circulator. Further discussion on the optical circulator will be reserved for Section 8.5.

The other subsystems that can be integrated are the RF downconverter, waveform generator, and photonic ADC. The RF downconverter depends on the modulator, amplifier, filter, and photodetector. For a parallel implementation of the downconverter, the advantage of the integrated version is the environmental stability that comes from miniaturization onto a common platform. In addition, the downconverter can be integrated with either an electronic- or photonic-based oscillator on the same chip. As for the waveform generator and the ADC, they both share the need for a pulsed laser source. The use of an integrated ring resonator for generating an optical pulse stream has been demonstrated by multiple groups [32, 33]. For the ADC, an IPC can provide all the required parts. For example, the system shown in Figure 7.6 can be realized. The electronics required for the clock recovery, and the array of electronic ADCs, can be combined with the optical modulator, switch, and array of photodetectors to realize the full subsystem. For the waveform generator, the one piece that is missing is the SLM. Fortunately, this function can be accomplished by a wavelength demultiplexer along with a phase and amplitude control piece. This is similar to the FIR filter block shown in Figure 5.6.

8.4 Other Applications in IPCs

The demonstration architecture utilizes a single antenna for both reception and transmission of signals at the same time. A single RF aperture is used to show how the RF photonic circulator can separate the bidirectional signal paths. However, some modern RF systems do not use a single antenna. Instead an array of antennas is used. One of the advantages of such a configuration is the overall RF gain for this type of aperture. An array of antennas can have *N* times the gain of a single antenna, where *N* is the number of elements in the array. For the increase to be realized, the signals from each of the elements of the array have to add coherently. This is known as beamforming the received signal. In order to beamform the signal, the phase of the signal from each element has to be controlled independently. In addition, to avoid unwanted grating lobes in the antenna pattern, the spacing between elements has to be no more than half a wavelength of the highest frequency that will be received. For signals above 30 GHz, the spacing can be on the order of millimeters. Placing a discrete photonic modulator at each element will not physically fit in such a tight spacing. An IPC-based modulator should be able to fit. In addition, the IPC can have a low-noise amplifier integrated between the antenna element and the modulator. Making multiple copies of the photonic circuit will allow each element to be upconverted into the optical domain.

In order to address the need for phase control, different methods can be implemented. An electronic phase shifter can be integrated into each of the photonic circuits behind each element. Unfortunately, the electrical phase shifter will only provide the correct phase shift for a small bandwidth of signals. Since the system is meant to be wideband, another method must be used. A photonic-based phase shifter has been shown to provide a flat phase shift over a large bandwidth of frequencies [34]. The phase change can be accomplished with a change in RF power of less than 0.2 dB over the entire frequency range. In Figure 8.4, the phase of a set of signals over 4 GHz of bandwidth can be tuned from +90° to –90°. Another method to ensure that the signals add coherently is to use a true time-delay element. In true time delay, the physical path that the signal travels over is changed. Often this

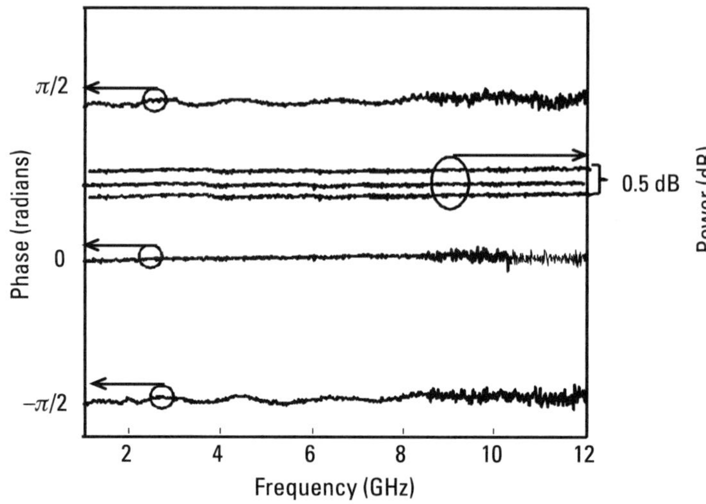

Figure 8.4 Phase shift from 90° to −90° degrees over a 1–12-GHz operating frequency range along with power change over the entire range of phase change.

involves the use of delay lines that can be switched in and out in order to get the correct amount of delay. While this type of delay line is very short, implementing them in an electronic circuit can be very lossy and require amplifiers, which, in turn, add noise. A photonic circuit version would be able to accomplish the same task but with less loss and potentially without the need for additional amplification. Figure 8.5 shows an example of a delay-line architecture along with switches.

With phase control, the array of antennas can now be steered to beamform signals that arrive at a certain angle to the front of the array. When the phase of each of the elements of the array is set correctly, a signal received from a specified angle will constructively add, while signals arriving from other directions will not. By changing the phases at each element, the array can be steered to different receive angles. However, the array can only be steered to one angle at a time. In order to form more than one beam at the same time, multiple phase shifters are needed at each element. The signal from each element would have to then be split to the multiple phase shifters and then recombined along separate paths. Such a configuration would be quite large. One of the advantages of using photonics is the ability to beamform multiple signals at the same time. Instead of using one laser, multiple lasers at different wavelengths can be combined through a wavelength multiplexer into the same modulator. Then each wavelength will pass through the same photonic delay lines. Due to dispersion, each of the wavelengths will see a different time delay. The different time delays mean that each wavelength

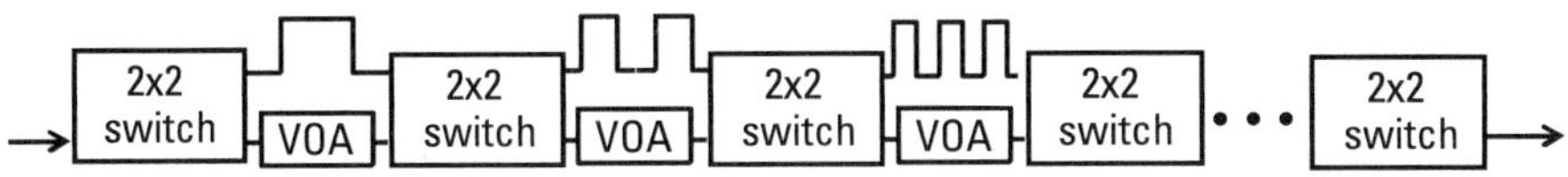

Figure 8.5 Block diagram of a true time-delay circuit using switches for variable delays.

will beamform a different angle from the array. Thus, the photonic circuit can form multiple beams at the same time.

8.5 Further Work in IPCs for Analog Applications

IPC technology has reached the point that it can now be considered to address the needs of RF photonic subsystems. However, a couple of points still need to be addressed. The first is the lack of an integrated optical circulator and isolator. Note that an optical isolator is simply an optical circulator with the third port terminated, so this discussion will cover both components. Discrete optical circulators depend on a ferromagnetic material in order to isolate the three ports. Integration of ferromagnetic materials onto an integrated platform has proven to be troublesome. Work has been focused on ferromagnetic materials such as cerium ytterbium garnet (Ce:YiG) to try and integrate with silicon platforms for isolators [35]. Recent work has shown integrated devices that can provide isolation as high as 20 dB [36] as well as dynamically tunable circulators [37]. While these demonstrations are promising, they have to be made repeatable enough to be transferred into a foundry process for high-yield fabrication.

This leads to the second point. For all the work that has been covered in IPCs, there is a limited amount of access for designing and fabricating these chips in small quantities. The electronic industry addressed this problem with services such as the metal oxide semiconductor implementation service (MOSIS) [38]. The service allows for multiple users to design into a shared wafer. These multiproject wafers (MPWs) allow users to fabricate prototype circuits without having to buy an entire wafer. The resulting cost reduction allows groups such as universities and small businesses access to foundries. The IPC area needs a similar service. The European Union is addressing this need with the Joint European Platform for Photonic Integration of Components and Circuits (JePPIX) [39]. JePPIX is looking to bring together indium phosphide and silicon nitride platforms together under one service. MPWs would be available for access to members. The ePIXfab service is also a consortium providing silicon photonic design access and contracting out the fabrication without actually having its own foundry (known as a fabless service) [40].

Also, the United States recently set up a manufacturing institute in IPCs called AIM Photonics [41]. While both services are looking to integrate both traditional electronic circuits along with photonic circuits, the main customers appear to be digital data communications. However, just as RF photonics took the digital communications discrete components and refined them for analog applications, the RF IPCs may also have to leverage the larger market provided by digital data applications.

8.6 Conclusions

IPCs have the potential to meet the needs of current and future analog applications. The current advantage of these circuits is a reduction in size and isolation from environmental fluctuations. By heterogeneously integrating different material structures onto silicon waveguides, the best performance can be achieved. This chapter reviews the various components that are required for IPCs as well as the subsystems that could be realized. While demonstrations have been made that show the promise of these photonic circuits, they have not been made accessible to general users. The AIM Photonics institute appears to be addressing this issue in the United States. Much like the evolution of discrete components, IPCs will most likely be pushed forward by digital communication requirements first. The use of photonic circuits for communications between mainframes in large server farms may lead to a high-yield silicon photonic process. In turn, this will reduce the cost of these components and allow for a reduction in costs for the analog version of these components, along with further advancements. In the end, the same path as discrete photonic development may be followed for IPCs. Namely, digital applications will allow for analog applications to be realized at a cheaper cost. However, improvements in performance will still need to be realized.

References

[1] Devgan, P., "Signal Processing Utilizing Radio Frequency Photonics," *AFRL Report,* AFRL-RY-WP-TR-2017-0172, 2017.

[2] Irmscher, S., R. Lewen, and U. Eriksson, "InP-InGaAsP High-Speed Traveling-Wave Electroabsorption Modulators with Integrated Termination Resistors," *Photonics Technology Letters,* Vol. 14, No. 7, 2002, pp. 923–925.

[3] Bach, H. G., et al., "InP-Based Waveguide-Integrated Photodetector with 100-GHz Bandwidth," *Journal of Selected Topics in Quantum Electronics,* Vol. 10, No. 4, 2004, pp. 668–672.

[4] Bowers, J., "Heterogeneous silicon III–V Photonic Integrated Circuits," *Proc. of Optical Fiber Communications Conference and Exhibition,* Los Angeles, CA, 2015.

[5] Boss, M., D. Evans, and E. Jacobs, "Performance of a 40-GHz RF Photonic Balanced Link Using a Polarization Modulator," *Proc. of Avionics, Fiber-Optics and Photonics Technology Conference,* San Diego, CA, 2013.

[6] Famà, S., et al., "High Performance Germanium-on-Silicon Detectors for Optical Communications," *Applied Physical Letters,* Vol. 81, No. 4, 2002, p. 586.

[7] Boyraz, O., and B. Jalali, "Demonstration of a Silicon Raman Laser," *Optics Express,* Vol. 12, No. 21, 2004, pp. 5269–5273.

[8] Liu, A., et al., "High-Speed Optical Modulation-based on Carrier Depletion in A Silicon Waveguide," *Optics Express,* Vol. 15, No. 2, 2007, pp. 660–668.

[9] Timurdogan, E., et al., "An Ultralow Power Athermal Silicon Modulator," *Nature Communications,* Vol. 5, No. 4008, 2014, pp. 1–11.

[10] Camacho-Aguilera, R. E., et al., "An Electrically Pumped Germanium Laser," *Optics Express,* Vol. 20, No. 10, 2012, pp. 11316–11320.

[11] Liang, D., and J. E. Bowers, "Recent Progress in Lasers on Silicon," *Nature Photonics,* Vol. 4, No. 8, 2010, pp. 511–517.

[12] Roelkens, G., et al., "Adhesive Bonding of InP/InGaAsP Dies to Processed Silicon-on-Insulator Wafers Using DVS-bis-benzocyclobutene," *Journal of the Electrochemical Society,* Vol. 153, No. 12, 2006, pp. G1015–G1019.

[13] Kostrzewa, M., et al., "InP Dies Transferred onto Silicon Substrate for Optical Interconnects," application. *Sensors and Actuators A: Physical,* Vol. 125, No. 2, 2006, p. 411–414.

[14] Toney, J., *Lithium Niobate Photonics,* Norwood, MA: Artech House, 2015.

[15] Guarino, A., et al., "Electro–optically Tunable Microring Resonators in Lithium Niobate," *Nature Photonics,* Vol. 1, No. 7, 2007, pp. 407–410.

[16] Rao, A., et al., "High-Performance and Linear Thin-Film Lithium Niobate Mach Zehnder Modulators on Silicon up to 50 GHz," *Optics Letters,* Vol. 41, No. 24, 2016, pp. 5700–5703.

[17] Kik, P. G., and Polman, A., "Erbium-Doped Optical-Waveguide Amplifiers on Silicon," *MRS Bulletin,* Vol. 23, No. 4, 1998, pp. 48–54.

[18] Rabus, D. G., *Integrated Ring Resonators,* Berlin Heidelberg: Springer-Verlag, 2007.

[19] Little, B. E., et al., "Ultra-Compact Si/SiO2 Microring Resonator Optical Channel Dropping Filters," *Photonics Technology Letters,* Vol. 10, No. 4, 1998, pp. 549–551.

[20] Xu, Q., et al., "Cascaded Silicon Micro-Ring Modulators for WDM Optical Interconnection," *Optics Express,* Vol. 14, No. 20, 2006, pp. 9431–9436.

[21] Chen, L., N. Sherwood-Droz, and M. Lipson, "Compact Bandwidth-Tunable Microring Resonators," *Optics Letters,* Vol. 32, No. 22, 2007, pp. 3361–3363.

[22] De Cort, W., et al., "Tuning of Silicon-on-Insulator Ring Resonators with Liquid Crystal Cladding Using the Longitudinal Field Component," *Optics Letters,* Vol. 34, No. 13, 2009, pp. 2054–2056.

[23] Li, Q., et al., "High-Power Flip-Chip Bonded Photodiode with 110 GHz Bandwidth," *Journal of Lightwave Technology,* Vol. 34, No. 9, 2016, pp. 2139–2144.

[24] Beling, A., and J. C. Campbell, "InP-based High-Speed Photodetectors," *Journal of Lightwave Technology,* Vol. 27, No. 3, 2009, pp. 343–355.

[25] Shi, Jin-Wei, et al., "Extremely High Saturation Current-Bandwidth Product Performance of a Near-Ballistic Uni-Traveling-Carrier Photodiode with a Flip-Chip Bonding Structure," *Journal of Quantum Electronics,* Vol. 46, No. 1, 2010, pp. 80–86.

[26] Arakawa, Y., et al., "Silicon Photonics for Next Generation System Integration Platform," *IEEE Communications Magazine,* Vol. 51, No. 3, 2013, pp. 72–77.

[27] Parekh, M. S., P. A. Thadesar, and M. S. Bakir, "Electrical, Optical And Fluidic Through-Silicon Vias for Silicon Interposer Applications," *Proc. of Electronic Components and Technology Conference,* 2011.

[28] Killge S., et al., "Optical Through-Silicon Vias," in *3D Stacked Chips* (I. Elfadel, G. Fettweis, eds.), New York, NY: Springer, 2016 pp. 21–234.

[29] Trinh, P. D., et al., "Silicon-on-Insulator (SOI) Phased-Array Wavelength Multi/Demultiplexer With Extremely Low-Polarization Sensitivity," *Photonics Technology Letters,* Vol. 9, No. 7, 1997, pp. 940–942.

[30] Chen, H.-W., et al., "Integrated Microwave Photonic Filter on a Hybrid Silicon Platform," *Transactions on Microwave Theory and Techniques,* Vol. 58, No. 11, 2010, pp. 3213–3219.

[31] Matsko, A., et al., "Whispering Gallery Mode-based Optoelectronic Microwave Oscillator," *Journal of Modern Optics,* Vol. 50, No. 15-17, 2003, pp. 2523–2542.

[32] Kippenberg, T. J., R. Holzwarth, and S. A. Diddams. "Microresonator-Based Optical Frequency Combs," *Science,* Vol. 332, No. 6029, 2011, pp. 555–559.

[33] Ferdous, F., et al., "Spectral Line-By-Line Pulse Shaping of On-Chip Microresonator Frequency Combs," *Nature Photonics,* Vol. 5, No. 12, 2011, pp. 770–776.

[34] Chan, E. H. W., and R. A. Minasian, "Photonic RF Phase Shifter and Tunable Photonic RF Notch Filter," *Journal of Lightwave Technology,* Vol. 24, No. , 2006, pp. 2676–2682.

[35] Bi, L., et al., "On-Chip Optical Isolation in Monolithically Integrated Nonreciprocal Optical Resonators," *Nature Photonics,* Vol. 5, No. 12, 2011, pp. 758–762.

[36] Tien, M. C., et al., "Silicon Ring Isolators with Bonded Nonreciprocal Magneto-Optic Garnets," *Optics Express,* Vol. 19, No. 12, 2011, pp. 11740–11745.

[37] Huang, D., et al., "Dynamically Reconfigurable Integrated Optical Circulators," *Optica,* Vol. 4, No. 1, 2017, pp. 23–30.

[38] https://www.mosis.com/what-is-mosis .

[39] http://www.jeppix.eu/welcome/.

[40] http://epixfab.eu/about/about-epixfab/.

[41] http://www.aimphotonics.com/.

9

Conclusions

The field of RF photonics has progressed significantly in the past 20 years. After the explosion in the use of photonics for digital communications in the late 1980s to late 1990s, a strong push was made to incorporate photonics into analog applications, such as cable TV distribution. Unfortunately, the same photonic components that were able to provide longer reach and higher bandwidth for digital applications were not able to compete with electronic solutions in RF applications. The high RF loss and NF associated with early RF photonic links kept them from being widely used. The optical fiber had already reached its optimal loss performance of just under 0.2 dB/km, while the optical amplifier could provide significant optical gain with an associated optical NF around 3 dB. On the other hand, the laser, modulator, and photodiode were identified as the weak components for use in analog applications. Most digital networks are designed to avoid nonlinear effects that occur when a significant amount of optical power propagates down tens of kilometers of optical fiber. Since the networks use wavelength-division multiplexing, each individual laser's power is kept at 1 mW or less so that the aggregate optical power does not exceed the nonlinear threshold. In addition, the modulator for digital applications is not optimized for linearity or insertion loss, as the digital data can be corrected in postprocessing and the optical power is purposefully kept low. Finally, the low power of the laser reduces the requirement on the power-handling of the photodiode.

9.1 A Brief Review of RF Photonics

Having identified the issue, these three components—the laser, the modulator, and the photodiode—were the subject of a large amount of research and engineering aiming to improve their performance. The RF photonic requirements for these components are more stringent, requiring advancements in their performance. Starting with the laser, the RIN has been improved significantly. Where a RIN of –145 dBc/Hz would be acceptable for digital applications, RIN on the order of –155 dBc/Hz is now realizable in commercially available parts [1] with –165 dBc/Hz available from specially designed devices [2]. Along with these noise results, the output power needs to be maximized. Commercially available lasers at 100 mWs of output power in the 1,550-nm range can achieve the above noise levels, while research has produced lasers with higher output powers available with low noise. Finally, the laser linewidth needs to be minimized. While not as important in the intensity-modulated direct detection link used in the demonstration architecture, the linewidth does impact potential use with phase modulation. Commercially available lasers with linewidths on the order of 1 kHz are available [3] while specialty lasers can achieve subhertz linewidths [4]. However, these types of lasers are only available at 20 mW and do not have the lowest RIN. Often these trade-offs, along with others, play a role in the choice of modulation format.

After the laser, the modulator is a key component. Extending the RF bandwidth is necessary as digital communications has not gone beyond 40 Gbps on one polarization of a single optical carrier. Commercially available phase modulators at 100 GHz are now available [5, 6], with Mach Zehnder interferometer-based intensity modulators also being available at similar frequencies. Research work showing modulation out to 300 GHz [7] has also been demonstrated. Along with the extended frequency operation, the electro-optic efficiency (directly related to the V_π) and the insertion loss has to be optimized. Specialty devices have been shown with low V_π [8] and an insertion loss of around 3 dB.

Finally, the photodetector continues to have its performance increased. Like the modulator, the extended bandwidth is necessary to meet the requirements of the RF photonic systems. Commercially available components out to 20 GHz are readily available with high responsivity (optoelectronic efficiency) [9]. Achieving higher power-handling and higher frequency has required new designs. While research has demonstrated devices with both high power and ultrawide frequency [10, 11], commercially available devices are only now starting to compete [12]. Again, note that the high power and ultrawide frequency often come at the expense of reduced responsivity.

While the optical amplifier is also an important component, the performance has already been optimized. The erbium-doped fiber amplifier already has an optical NF near the theoretical limit of 3 dB, and the noise penalty can be optimized. In terms of compactness, the EDFA can be made in very small packages [13]. While amplifiers using semiconductor or nonlinear optical effects continue to be explored, the EDFA is still the chosen amplifier.

The optical filter has been well explored in terms of the use of a FBG. Commercially available devices with bandwidths on the order of 50 MHz are available [14]. While their tunability is limited, their optical insertion loss and their high Q, makes them preferred devices for RF applications.

9.2 Discrete-Based RF Photonic Subsystems

The transmission of analog signals over a fiber-optic link is one of the first and most basic applications of RF photonics. As described in Chapter 2, there are multiple ways to realize an RF photonic point-to-point link. Commercial vendors provide both MZM-based photonic links [15, 16] and direct laser modulation-based ones [17]. An extension of a simple RF photonic link, vendors also provide analog delay lines, or analog memory units [18]. Due to a need for higher-fidelity delay lines, research work continues to seek improvement in the RF performance of the photonic delay lines [19]. The demonstrated performance allows for delay lines that can operate up to 40 GHz with a dynamic range on the order of 103-107 dB•Hz$^{2/3}$ [20], while other delay lines have been demonstrated with a NF less than 10 dB and with a positive gain of almost 40 dB [21]. Clearly, delay lines can be made with different applications in mind.

Appearing just after the antenna, the RF photonic-based circulator is a subsystem that serves multiple purposes. It isolates the transmission and reception path, allowing the antenna to serve two functions at once. The circulator also takes the received signals and upconverts them into the optical domain. By doing so, the signals can now be operated on by other photonic-based systems. Work in photonic-based RF circulators has focused on different needs. In some cases, the emphasis is on a single antenna aperture, isolated from other sources of interference. Here, the circulator has to handle very high RF powers on the transmission side. Also, the received signals can be very low-power, requiring the circulator to have a low NF. In other cases, the antenna could be part of a larger array of antennas, reducing the required power at each aperture. However, the various antennas can couple in strong unwanted signals. Under this condition, the circulator has to be able to operate over a large range of input power without having intermodulation distortion products appearing in the receive path. The SFDR of the circulator is the important RF metric for this application. Research work continues on these RF photonic-based circulators [22] with possible paths to providing commercial products [23].

The signal separation subsystem depends strongly on optical filter technology. Commercial vendors are strong in both FBG [24] as well as in thin film-based filters [25]. AWGs are another optical filter technology that can be used for signal separation, with at least one commercial vendor providing solutions [26]. With the use of an AWG, the finite impulse response filter can be realized. In order to overcome the coherence problems, a multiwavelength source has to be used. One method is to use a bank of lasers, each operating at their own independent wavelength. Banks of lasers are available commercially [27], while research on a single source that can provide multiple independent optical carriers is ongoing [28]. The infinite impulse response filter can also be realized by the multimode OEO [29].

The photonic RF oscillator has been investigated thoroughly. The OEO has been around since its invention in 1996. Commercial OEO systems are now also available for purchase from commercial vendors [30]. Nevertheless, research work continues on OEOs, with one of the main areas focused on increasing the tuning range of the frequency of the signal from the OEO [31].The other photonic oscillator highlighted in Chapter 8 is based on the optical injection locking of two lasers. Research work in this area is ongoing with hybrid systems that combine the OEO feedback with the flexibility of the optical injection locking scheme showing promise [32]. Beyond these two types of photonic RF oscillators, optical frequency combs can also generate low phase noise RF signals. In fact, the carrier envelope phase-locked optical frequency comb source can have very low phase noise [33]. The system takes advantage of a mode-locked laser to generate very short optical pulses. The optical spectrum associated with these very short pulses can cover over an octave in frequency. By taking a spectral comb line, frequency-doubling it, and then beating the doubled signal with the spectral comb line of the same frequency, an error signal can be generated. The error signal can then be used to correct the envelope offset. The result is a very stable frequency source that can be used as an RF clock source. The carrier envelope phase-locked frequency comb source is now available for purchase through multiple vendors [34, 35].

The signal identification block depends on a form of spectrum analysis. The use of SHB in a crystal can capture a wideband RF spectrum [36]. The RF spectrum is upconverted onto an optical carrier and then written into the rare Earth-doped crystal. A second laser is then used to read out the stored RF spectrum, allowing the center frequency of any signal within the bandwidth to be recovered. A commercial company is developing a crystal-based spectrum analyzer for both frequency identification and analog data processing [37]. Research work in methods using dispersion for frequency identification is also ongoing. One other method involves the use of a recirculating loop [38]. An RF signal can be input to a long fiber loop with dispersion. After each trip through the loop, the RF sidebands will see different time delays. When enough delay is accumulated, the time-delayed sidebands are autocorrelated at a detector. The autocorrelation signal is then processed and the center frequency is determined.

Another method for spectrum analysis is channelization. A channelizer takes a wideband RF spectrum and decimates it into smaller parts or channels. Each channel can be processed in parallel. RF photonics lends itself to channelization by the use of optical filters that can be arrayed to split the wideband spectrum into the individual channels. An example of an RF photonic-based channelizer has been previously demonstrated [39]. The RF spectrum is modulated onto an optical carrier. The upconverted optical signals are then passed through an optical filter bank. After the filter bank, each of the channels is then downconverted to an intermediate frequency and digitized for processing. Many examples of RF photonic channelizers have been investigated, including using an optical frequency comb [40] as well as a parallel set of chirped filters that chops up a larger spectral bandwidth to smaller 2-GHz channels [41].

The processing block of the demonstration architecture contains both the downconversion subsystem and the ADC. Both of these technologies continue to be explored in multiple architectures. Downconversion techniques are now incorporating multiple sidebands in order to spectrally fold higher-order Nyquist bands to the first Nyquist band [42]. This method is similar to the electronic Nyquist folding receiver [43]. Other methods have used optical pulses in order to downconvert higher-frequency RF signals [44]. New methods continue to be explored, taking advantage of new advancements in optical combs as well as short pulses.

The use of stable optical pulses in photonic ADCs is also pushing the performance of these systems. While a large amount of work is still being done on improving sampling jitter in the area of research [45], there have been demonstrations of photonic sampling being used in real-world systems [46]. However, photonic-based analog-to-digital conversion is still based on discrete components. Both analog-to-digital and downconverter systems will benefit from IPC advancements, perhaps finally allowing them to compete with purely electronic solutions.

9.3 Alternative Systems Using RF Photonics

The demonstration architecture presented throughout the book is just one example in which different RF photonics subsystems can be used. Multiple other examples exist of system architectures that incorporate different types of RF photonic subsystems. RF photonic links are the most commonly used for different applications. In addition to their use in antenna remoting and analog delay lines, RF photonic links can be used for other applications. One example is the use of an RF photonic link for in-situ antenna calibration [47]. A large antenna array often needs to be calibrated after long operational periods. To provide a calibration signal to each of the antenna elements, long copper coax cables are used. As these metal cables block or distort any incoming signals, the antenna system has to be taken out of service while the calibration is done. However, a fiber-optic cable can be used to provide the RF signal to the antenna elements without interfering with any incoming signal. The key element in this setup is an unbiased photodiode at end of each fiber-optic link. The photodiode can convert the optical signal back to the RF domain to excite the antenna. Since the photodiode is unbiased, it can only handle a certain amount of optical power, and the RF conversion is very lossy. Fortunately, the photodiode can be placed very close to the antenna element, compensating for the large loss. Investigations into the nonlinear performance of the unbiased photodiode and its utility for antenna calibration have been done [48, 49]. While not a traditional use of an RF photonic link, it is one that makes use of the components developed for RF photonic applications.

Another example is a photonically fed array of tightly coupled antenna elements. Tightly coupled arrays have been shown to have ultrawideband RF performance [50]. However, the elements are often closely spaced and require RF feeds that can operate over a large bandwidth. On top of this, the impedance of the tightly coupled antenna elements may not be easily matched with a traditional coaxial cable. Any significant mismatch in the impedance will result in reflections and reduce the radiated power from the antenna. To overcome these limitations, an RF photonic feed has been demonstrated with a tightly coupled array [51]. A fiber-optic cable is connected to a photodiode behind each of the elements. The photodiodes are biased in this case, as opposed to the antenna calibration example presented above. The photodiode converts the optical signal into an RF signal that will then be radiated out of the antenna. The photodiode has very low resistance, allowing an external impedance to be used to match the impedance of the element. Previous demonstrations used a ultrawide frequency photodiode connected to single antenna to send digital data on a 120-GHz RF carrier [52]. Using an array ultrawide frequency photodiode with moderately high RF output can make use of the ultrawideband capability of the tightly coupled array [53]. A widely tunable RF signal can be generated in the photonic domain, by beating two optical carriers at the photodiode. Along with the wideband capability, a photonically fed antenna can provide the advantages of lower weight and size, while also providing immunity to electromagnetic interference.

While the previous two examples focused on radiating RF signals over large areas or distances, the same technology can be used for much smaller scale applications. A photoconductive probe can be used to perform time domain reflectometry on integrated electronic circuits [54]. As the circuits' node size becomes smaller and smaller, small defects can reduce the performance. To detect these small defects, very short time transitions are needed. The advantage of using RF photonics is the availability of short optical pulses. The short optical pulses can resolve neighboring defects with a ten times improvement when compared to electronic time domain reflectometry. The use of specially designed photoconductive probes provides advantages over other demonstrations [55]. To measure the circuit under test, the optical pulses are used as an input to the injection photoconductive probe. The resulting terahertz radiation is passed out the probe tip into the test circuit. Then, a detection probe tip is used. Using a copy of the optical pulses allows for the coherent detection of the radiation at the detection probe tip. The detection probe will sample any reflections from circuit defects. Along similar lines, a ultrawide frequency photodiode can be used to radiate millimeter or terahertz radiation through a sample [56]. Note that the sample can be an integrated circuit or a biological sample. The input to the photodiode is often generated by the beating of two laser sources at the photodiode. After passing through the sample, the radiation is mixed down with a LO, also generated by the beating of two lasers. A lock-in amplifier is used to recover both amplitude and phase information. The key components of this system are an oscillator based on photonic sources and a ultrawide frequency photodiode.

While many other RF photonic link demonstrations exist, other systems have also been demonstrated. An RF photonic-based coherent radar system has been developed and used in a real-world situation [57]. At the heart of the photonic radar system is a low-jitter optical pulse source. The pulse source can be used to create different waveforms for transmission from the radar. The carrier frequency of the photonically generated can as high as 40 GHz, without the need for RF upconversion. On the receive side of the radar, the same optical pulse source can be used as the optical sampler for a photonic ADC. Again, the system requires no downconversion on the receive end. The system has even been used in real-world environments in order to track commercial airplanes. The track data from the photonic radar agreed very well with other data sources, confirming that the radar operated as expected. When compared to a traditional electronic-based radar system, the main advantages of the photonic-based system are higher carrier frequency and lower timing jitter. The main weakness in the photonic system is a lower dynamic range on the receiver as compared to the electronic equivalent. A review of the radar performance has been performed independently [58]. The review article notes that the system may be limited in the range over which the radar can accurately operate. Improvements in the long-term stability of the optical pulse source can lead to better range performance. In spite of these limitations, the article also notes that the radar demonstration shows great promise in leading to a larger acceptance of RF photonic-based systems in real-world applications.

Other applications of RF photonics have been pursued using subsystems in nontraditional ways. An OEO can be driven into a chaotic state [59]. The optoelectronic oscillator is similar to the MM-OEO demonstrated in Chapter 5. However, where the MM-OEO is kept below threshold, the chaos OEO is set to operate above threshold. The experimental and theoretical work was able to map the different operating states of the OEO. When the OEO was driven into a specific chaotic state, a flat, broadband power spectrum was generated. As noted in the article, the chaotic behavior of the OEO is different than other chaotic systems. This investigation could lead to fundamentally new observations in chaos.

RF photonic systems can also perform Fourier processing on signals without the need for digitization. The use of optics for Fourier transformation and processing is well understood [60]. Arbitrary waveform generation is one example of an RF photonic application that makes use of an optical Fourier transform. While an early demonstration of waveform generation used a free-space setup involving two lenses and a spatial mask [61], more recent examples use other methods. Another example is the rare Earth-doped crystal-based spectrum analyzer. The spectrum analyzer takes the time domain signals and stores the energy in the frequency domain. The Fourier transform allows the spectrum analyzer to capture wideband signals instantaneously [62]. Finally demonstrations have shown the capability to do fractional Fourier transforms (FrFTs) on RF signals [63]. The FrFT can be useful for recovering information on chirped RF waveforms. Performing this operation in the photonic domain reduces the latency associated with this type of processing.

Finally RF photonics can be used for advanced sensors. Two examples are a millimeter-wave radiometer and photonic-based gyroscopes. A radiometer is often used to measure the amount of radiation that is emitted from a blackbody source [64]. A basic radiometer consists of an antenna, a detector, and an integrator. A photonic implementation of a radiometer allows the system to measure noise powers at millimeter-wave frequencies or higher, beyond what is capable of using purely electronic means [65, 66]. In fact, such a system could be used as a passive imager [67]. At least one commercial company has even begun to use such imaging technologies for different applications [68].

As a second example, a gyroscope is used to measure the angular velocity of an object. The photonic gyroscope has been used for many years in many different applications [69]. The photonic implementation of the gyroscope has provided higher sensitivity and stability than other demonstrations. As the limitations on the performance of these gyroscopes have been investigated, the need for high-power-handling photodiodes, low-noise lasers, and low-loss optical loops has been identified [69]. All of these components are being developed to meet other RF photonic needs. The performance of photonic gyroscopes will increase as the RF photonic components improve. Thus, photonic gyroscopes can also be considered a part of the RF photonic application space. Clearly, many examples of RF photonic applications exist. While many are covered here, the list is by no means complete. Those that are left out are not due to any technical deficiency but merely limited space.

9.4 Future Work in RF Photonics

New applications will continue to be developed as both the needs of future systems increase in frequency and instantaneous bandwidth and the component performance improves. Wireless communications has become one of the new drivers for technological innovations. With the introduction of the smart phone, wireless data usage continues to increase. As opposed to traditional long-haul communications, where each household needs access to high-speed data, wireless communications provides individual users access to the same amount of data bandwidth. In order to meet this demand, wireless providers now need to utilize higher-frequency carriers. Fifth-generation systems are already considering the use of carrier frequencies in the 30-GHz or higher range in order to meet the data demand, as well as to access a part of the spectrum that is less congested. Other commercial and military applications are also looking to exploit the millimeter-wave region of the spectrum.

Looking to the near future of RF photonics, three key developments appear to be the driving factors in new discrete component-based RF photonic applications: the optical frequency comb, the high-Q optical cavity, and the high-power, ultrawide frequency photodiode. The optical fiber and the amplifier used in discrete systems are mature and already are well near the practical limits in terms of performance. While electro-optic modulator performance has room for improvement, the electro-optic conversion efficiency trades RF gain and NF improvements for dynamic range or vice versa. Besides, many electronic and photonic means exist for improving the overall performance of the modulator without changing the electro-optic conversion efficiency.

Advancements in the optical frequency comb have utility in many of the RF photonic applications previously discussed. In the same way that digital photonic networks make use of WDM with multiple optical carriers, RF photonics will benefit from optical frequency combs. From oscillators to downconversion to waveform generation to ADCs, all of these can benefit from the optical frequency comb. The ability to replace multiple lasers with a single source that can generate coherent optical lines is not only cost-saving, but can lead to new signal-processing techniques.

The high-Q photonic cavity will allow for many different applications. The oscillator, signal separation, and signal identification systems can make use of a high-Q cavity. An OEO with a high-Q photonic cavity will provide low-phase noise RF sources without the need for long fiber lengths. The removal of the long fiber cavity will reduce the sensitivity of the system to environmental fluctuations and increase the nonoscillating sidemode frequency spacing. The high-Q cavity will also provide higher resolution for spectrum analysis, channelization, and the identification of the center frequency of RF signals.

Finally, the ultrawide frequency, high-power photodiode will benefit the transmission of RF signals. As mentioned above, wireless technologies will need to be able to transmit higher-frequency RF signals. The ability to feed an antenna with a higher-frequency signal at the top of a tall cellular tower will require photonics. By placing a photodiode as the feed to the antenna, the system will be able to accommodate these signals. Also, the ultrawide frequency photodiodes will also be able to nondestructively probe integrated circuits as well as biological samples.

The far future of RF photonics rests in IPCs. The reduction in size will allow these subsystems to become more ubiquitous—and less expensive. IPCs are also finally beginning to break out to more real-world uses. Multiple foundries exist around the world where a user can design a circuit using standard process design kits (PDKs). In fact, software for designing into multiple foundries that use both silicon and semiconductor materials is available [70]. The ability to design into these various foundries opens new opportunities for IPCs.

The application space for RF photonics will continue to grow in ways that cannot be completely predicted. What is known for sure is the continued need to access higher frequencies and utilize larger bandwidths. RF photonics is well suited to address these needs and will continue to improve its performance and reliability. Along with continued work in IPCs, the next 20 years should see a large expansion in the use of RF photonic technologies.

References

[1] https://goochandhousego.com/product-categories/dfb-lasers-modules/.

[2] http://mortonphotonics.com/ultra_low_noise_lasers.

[3] http://www.rio-lasers.com/_products/planex.html.

[4] http://www.menlosystems.com/products/.

[5] http://www.eospace.com/.

[6] http://www.phasesensitiveinc.com/components/optical-modulators/.

[7] Macario, J., et al., "Full Spectrum Millimeter-Wave Modulation," *Optics Express,* Vol. 20, No. 21, 2012, pp. 23623–23629.

[8] Juodawlkis, P. W., et al., "InGaAsP/InP Quantum-Well Electrorefractive Modulators with Subvolt Vpi," *Proc. of SPIE,* Vol. 5435, 2004, pp. 53–63.

[9] http://www.discoverysemi.com/

[10] Li, Z., et al., "High-Power High-Linearity Flip-Chip Bonded Modified Unitraveling Carrier Photodiode," *Optics Express,* Vol. 19, No. 26, 2011, pp. B385–B390.

[11] Ito, H., et al., "High-Speed and High-Output InP-InGaAs Unitraveling-Carrier Photodiodes," *Journal of Selected Topics in Quantum Electronics,* Vol. 10, No. 4, 2004, pp. 709–727.

[12] https://www.finisar.com/optical-components

[13] https://www.nuphoton.com/product-category/amplifiers/pulsed-amplifiers/edfa-cw/.

[14] http://www.teraxion.com/en/tfn.

[15] http://www.pharad.com/rf-photonic-transceivers.html.

[16] http://www.photonicsinc.com/photonic_links.html.

[17] http://linphotonics.com/products/.

[18] http://www.b2bphotonics.com/products/.

[19] Diehl, J. F., et al., "Microwave Photonic Delay Line Signal Processing," *Applied Optics,* Vol. 54, No. 31, 2015, pp. F35–F41.

[20] Urick, V. J., et al., "Design and Performance Of Ka-Band Fiber-Optic Delay Lines," *NRL Memorandum Report,* NRL/MR/5650-12-9456, 2012.

[21] Singley, J. M., et al., "Design and Performance of a 560-microsecond Ku-Band Binary Fiber-Optic Delay Line," *NRL Memorandum Report,* NRL/MR/5650-15-9545, 2015.

[22] Yap, D., J. H. Schaffner, and O. M. Efimov. "Tunable Photonic RF Circulator for Simultaneous Transmit and Receive," United States Patent No. 9,250,452, 2016.

[23] http://www.photonicsinc.com/tirx.html.

[24] http://www.ofsoptics.com/fiber-bragg-gratings.html

[25] https://www.newport.com/c/fiber-optic-components.

[26] https://www.ntt-electronics.com/en/products/photonics/awg_mul_d.html.

[27] https://www.thorlabs.com/

[28] http://www.ramphotonics.com/products/.

[29] Devgan, P., V. J. Urick, and K. J. Williams, "Multimode Optoelectronic Oscillator," United States Patent 8,824,901, 2014.

[30] http://www.oewaves.com/products.

[31] Li, W., and J. Yao, "A Wideband Frequency Tunable Optoelectronic Oscillator Incorporating a Tunable Microwave Photonic Filter Based on Phase-Modulation to Intensity-Modulation Conversion Using a Phase-Shifted Fiber Bragg Grating," *Transactions on Microwave Theory and Techniques,* Vol. 60, No. 6, 2012, pp. 1735–1742.

[32] Suelzer, J. S., et al., "Tunable, Low-Phase-Noise Microwave Signals from an Optically Injected Semiconductor Laser with Opto-Electronic Feedback," *Optics Letters,* Vol. 42, No. 16, 2017, pp. 3181–3184.

[33] Jones, D., et al., "Carrier Envelope Phase Control of Femtosecond Mode-Locked Lasers and Direct Optical Frequency Synthesis," Physics Faculty Contributions, 86, 2000.

[34] http://www.menlosystems.com/.

[35] http://www.toptica.com/products/frequency-combs/.

[36] Schlottau, F., et al., "Spectral Hole Burning for Wideband, High-Resolution Radio-Frequency Spectrum Analysis," *Optics Letters,* Vol. 30, No. 22, 2005, pp. 3003–3005.

[37] https://www.s2corporation.com/.

[38] Stead, M. R., W. Zhou, and M.-C. Li, "Using Dispersion in a Fiber-Optic Loop to Perform Time Domain Analogue RF Signal Auto-Correlation," *Proc. of Optical Fiber Communication Conference and Exposition,* 2011.

[39] Rogge, M. S., V. J. Urick, and F. Bucholtz, "Analysis of an Optical Channelization Technique for Microwave Applications," *NRL Memorandum Report,* NRL/MR/5652-07-9061, 2007.

[40] Xie, X., et al., "Broadband Photonic Radio-Frequency Channelization Based on a 39-GHz Optical Frequency Comb," *Photonics Technology Letters,* Vol. 24, No. 8, 2012, pp. 661–663.

[41] Hunter, D. B., L. G. Edvell, and M. A. Englund. "Wideband Microwave Photonic Channelised Receiver," *Proc. of International Topical Meeting on Microwave Photonics,* 2005.

[42] Agarwal, A., et al., "Photonically-Enabled RF Front-End for Wideband Flexible Down-Conversion," *Optics Express,* Vol. 25, No. 7, 2017, pp. 7338–7348.

[43] Fudge, G. L., et al., "A Nyquist Folding Analog-to-Information Receiver," *Proc. of Signals, Systems and Computers,* 2008.

[44] McKinney, J. D., and V. J. Urick, "Radio-Frequency Down-Conversion Via Sampled Analog Optical Links," *NRL Memorandum Report,* NRL/MR/5650-10-9275, 2010.

[45] Khilo, A., et al., "Photonic ADC: Overcoming the Bottleneck of Electronic Jitter," *Optics Express,* Vol. 20 No. 4, 2012, pp. 4454–4469.

[46] www.dasphotonics.com.

[47] McDermitt, C. S., et al., "Performance of 16-Channel, Photonic, Phased-Array Antenna Calibration System," *Electronics Letters,* Vol. 45, No. 24, 2009, pp. 1249–1250.

[48] Godinez, M. E., et al., "RF Characterization of Zero-Biased Photodiodes," *Journal of Lightwave Technology,* Vol. 26, No. 24, 2008, pp. 3829–3834.

[49] Diehl, J. F., et al., "The Performance of a Fiber-Optic Link Using Unbiased Balanced Photodiodes for Antenna Array Calibration," *Transactions on Microwave Theory and Techniques,* Vol. 60, No. 1, 2012, pp. 195–200.

[50] Tzanidis, I., K. Sertel, and J. L. Volakis, "UWB Low-Profile Tightly Coupled Dipole Array with Integrated Balun and Edge Terminations," *Transactions on Antennas and Propagation,* Vol. 61, No. 6, 2013, pp. 3017–3025.

[51] Shi, S., et al., "Ultrawideband Optically Fed Tightly Coupled Phased Array," *Journal of Lightwave Technology,* Vol. 33, No. 23, 2015, pp. 4781–4790.

[52] Minotani, T., A. Hirata, and T. Nagatsuma, "A Broadband 120-GHz Schottky-Diode Receiver for 10-Gbit/s Wireless Links," *IEICE Transactions on Electronics,* Vol. 86 No. 8, 2003, pp. 1501–1505.

[53] Cross, A. S., et al., "High-Power Flip-Chip Mounted Photodiode Array," *Optics Express,* Vol. 21, No. 8, 2013, pp. 9967–9973.

[54] Nagel, M., A. Michalski, and H. Kurz, "Contact-Free Fault Location And Imaging with On-Chip Terahertz Time-Domain Reflectometry," *Optics Express,* Vol. 19, No. 13, 2011, pp. 12509–12514.

[55] Cai, Y., et al., "Electro Optical Terahertz Pulse Reflectometry—An Innovative Fault Isolation Tool," *Proc. of Electronic Components and Technology Conference,* 2010.

[56] Nagatsuma, T., et al., "Millimeter-Wave and Terahertz-Wave Applications Enabled By Photonics," *Journal of Quantum Electronics,* Vol. 52, No. 1, 2016, pp. 1–12.

[57] Ghelfi, P., et al., "A Fully Photonics-Based Coherent Radar System," *Nature,* Vol. 507, No. 7492, 2014, pp. 341–345.

[58] McKinney, J., "Technology: Photonics Illuminates the Future of Radar," *Nature,* Vol. 507, No. 7492, 2014, pp. 310–312.

[59] Callan, K. E., et al., "Broadband Chaos Generated by an Optoelectronic Oscillator," *Physical Review Letters,* Vol. 104, No. 11, 2010, pp. 113901–4.

[60] Goodman, J. W., *Introduction to Fourier optics,* Englewood, CO: Roberts and Company, 2005.

[61] Weiner, A. M., "Femtosecond Pulse Shaping Using Spatial Light Modulators," *Review of Scientific Instruments,* Vol. 71, No. 5, 2000, pp. 1929–1960.

[62] Gorju, G., et al., "10-GHz Bandwidth RF Spectral Analyzer with MHz Resolution Based on Spectral Hole Burning in Tm/sup 3+: YAG," *Photonics Technology Letters,* Vol. 17, No. 11, 2005, pp. 2385–2387.

[63] Schnébelin, C. and H. G. de Chatellus, "Agile Photonic Fractional Fourier Transformation of Optical and RF Signals," *Optica,* Vol. 4, No. 8, 2017, pp. 907910.

[64] www.millitech.com/pdfs/Radiometer.pdf.

[65] Zhang, Y., "Application of Millimeter-Wave Photonics Technology in Passive Millimeter-Wave Imaging," *Proc. of SPIE,* Vol. 7854, 2010, pp. 785436B1–785436B6.

[66] Schuetz, C. A., et al., "Radiometric Millimeter-Wave Detection Via Optical Upconversion and Carrier Suppression," *Transactions on Microwave Theory and Techniques,* Vol. 53, No. 5, 2005, pp. 1732–1738.

[67] Schuetz, C. A., et al., "Sparse Aperture Millimeter-Wave Imaging Using Optical Detection and Correlation Techniques," *Proc. of SPIE,* Vol. 6548, 2007, pp. 65480B1–65480B8, 2007.

[68] http://www.phasesensitiveinc.com/systems/video-rate-mmw-imager/.

[69] Ciminelli, C., et al., "Photonic Technologies For Angular Velocity Sensing," *Advances in Optics and Photonics,* Vol. 2, No. 3, 2010, pp. 370–404.

[70] http://www.phoenixbv.com/.

List of Acronyms

3R	reamplify, retime, and reshape
ACF	amplitude comparison function
ADC	analog-to-digital converter
AIM Photonics	American Institute for Manufacturing Photonics
AMP	amplifier
AMZ	asymmetric interferometer
AWG	arrayed-waveguide grating
BER	bit error rate
BERT	bit error rate tester
BPF	bandpass filter
CDR	compression dynamic range
CMOS	complementary metal oxide semiconductor
CW	continuous wave

dB	decibel
dBc	decibel relative to carrier
dBm	decibel relative to 1 mW
DC	direct current
DEMUX	demultiplexer
DET	detector
DFB	distributed feedback
DPMZM	dual-parallel Mach Zehnder modulator
EAM	electro-absorption modulator
EBF	electronic bandpass filter
EDFA	erbium-doped fiber amplifier
ENOB	effective number of bits
EOM	electro-optic modulator
FBG	fiber-Bragg grating
FEC	forward error correction
FIR	finite impulse response
FP	Fabry-Perot
FrFT	fractional Fourier transform
Gbps	gigabits per second
GRIN	gradient index
HF	high-frequency

IF	intermediate-frequency
IF AMP	intermediate-frequency amplifier
IFM	instantaneous-frequency measurement
IIR	infinite-impulse response
IMD_2	second-order intermodulation
IMD_3	third-order intermodulation
IMDD	intensity-modulated direct detection
JePPIX	Joint European Platform for Photonic Integration of Components and Circuits
km	kilometer
LiNbO3	lithium niobate
LO	local oscillator
LPF	low-pass filter
MDS	minimum detectable signal
MISF	maximum-input spur-free
MM-OEO	multimode optoelectronic oscillator
MOD	modulator
MOSIS	metal oxide semiconductor implementation service
MPW	multiproject wafer
MUX	multiplexer
mW	milliwatt

MZM	Mach-Zehnder modulator
NF	noise figure
OA	optical amplifier
OC-192	optical channel 192 (9.953 Gbps)
OEO	optoelectronic oscillator
PD	photodetector
PDK	process design kit
PIN	positive-intrinsic-negative
PN	positive-negative
PON	passive optical network
PSD	power spectral density
RF	radio frequency
RIN	relative intensity noise
SBS	stimulated Brillouin scattering
SFDR	spur-free dynamic range
SHB	spectral hole burning
SLM	spatial light modulator
SNR	signal-to-noise ratio
SONET	synchronous optical network
SSB	single-sideband
Tbps	terabits per second

TBPF	tunable bandpass filter
TFLN	thin-film lithium niobate
THz	terahertz
UHF	ultrahigh frequency
VCO	voltage-controlled oscillator
VHF	very high frequency
WDM	wavelength demultiplexer/division multiplex

About the Author

Preetpaul Devgan is currently a senior electrical engineer at the Air Force Research Laboratory (AFRL). Prior to joining AFRL, he worked at the Naval Research Laboratory, after receiving his Ph.D. from Northwestern University. Prior to that, he worked at Lucent Technologies in the optical networking group, which he joined after receiving his B.S. and M.S. from Purdue University. His current research interests include RF photonic systems for signal-processing applications, specifically optoelectronic oscillators, novel modulation formats, frequency conversion, and identification.

Dr. Devgan is a senior member of IEEE and the winner of the 2016 IEEE Dayton Section Photonics Society award. Previously, he received technology transfer awards from both the NRL and the AFRL Sensors directorate. Devgan attended Northwestern University under a National Science Foundation IGERT fellowship, where he was recognized as a finalist for the 2005 OSA New Focus/Bookham Student award. Prior to that, he held the office of president of the student chapter of Eta Kappa Nu at Purdue, during which time the chapter was recognized with the best student chapter award. Dr. Devgan has published over 60 peer-reviewed journal and conference publications (including multiple invited papers) and holds four patents.

Index

Recent Titles in the Artech House Applied Photonics Series

Brian Culshaw and Marc Wuilpart, Series Editors

The ABCs of Fiber Optic Communication, Sudhir Warier

Advanced Optical Communication Systems and Networks, Milorad Cvijetic and Ivan B. Djordjevic

Applications of Modern RF Photonics, Preetpaul Singh Devgan

Chemical and Biochemical Sensing with Optical Fibers and Waveguides, Gilbert Boisdé and Alan Harmer

Coherent and Nonlinear Lightwave Communications, Milorad Cvijetic

Coherent Doppler Wind Lidars in a Turbulent Atmosphere, Viktor Banakh and Igor Smalikho

Coherent Lightwave Communication Systems, Shiro Ryu

Digital Optical Measurement Techniques and Applications, Pramond Rastogi, Editor

DWDM Fundamentals, Components, and Applications, Jean-Pierre Laude

Fiber Bragg Gratings: Fundamentals and Applications in Telecommunications and Sensing, Andrea Othonos and Kyriacos Kalli

The Fiber-Optic Gyroscope, Second Edition, Hervé C. Lefèvre

Fiber-Optic Sensors for Biomedical Applications, Daniele Tosi and Guido Perrone

Finite Element Modeling Methods for Photonics, B. M. Azizur Rahman and Arti Agrawal

Frequency Stabilization of Semiconductor Laser Diodes, Tetsuhiko Ikegami, Shoichi Sudo, and Yoshihisa Sakai

Handbook of Distributed Feedback Laser Diodes, Second Edition, Geert Morthier and Patrick Vankwikelberge

Helmet-Mounted Displays and Sights, Mordekhai Velger

Introduction to Infrared and Electro-Optical Systems, Second Edition, Ronald G. Driggers, Melvin H. Friedman, and Jonathan Nichols

Introduction to Lightwave Communication Systems, Rajappa Papannareddy

Introduction to Radiometry and Photometry, Second Edition, William Ross McCluney

Introduction to Semiconductor Integrated Optics, Hans P. Zappe

Liquid Crystal Devices: Physics and Applications, Vladimir G. Chigrinov

Lithium Niobate Photonics, James E. Toney

Military Laser Technology and Systems, David H. Titterton

New Photonics Technologies for the Information Age: The Dream of Ubiquitous Services, Shoichi Sudo and Katsunari Okamoto, editors

Optical Document Security, Third Edition, Rudolf L. van Renesse

Optical FDM Network Technologies, Kiyoshi Nosu

Optical Fiber Amplifiers: Materials, Devices, and Applications, Shoichi Sudo, editor

Optical Fiber Communication Systems, Leonid Kazovsky, Sergio Benedetto and Alan Willner

Optical Fiber Sensors, Volume Three: Components and Subsystems, John Dakin and Brian Culshaw, editors

Optical Fiber Sensors, Volume Four: Applications, Analysis, and Future Trends, John Dakin and Brian Culshaw, editors

Optical Measurement Techniques and Applications, Pramod Rastogi

Optical Transmission Systems Engineering, Milorad Cvijetic

Optoelectronic Techniques for Microwave and Millimeter-Wave Engineering, William M. Robertson

Reliability and Degradation of III-V Optical Devices, Osamu Ueda

Signal Processing and Performance Analysis for Imaging Systems, S. Susan Young, Ronald G. Driggers, and Eddie L. Jacobs

Smart Structures and Materials, Brian Culshaw

Substrate Surface Preparation Handbook, Max Robertson

Surveillance and Reconnaissance Imaging Systems: Modeling and Performance Prediction, Jon C. Leachtenauer and Ronald G. Driggers

Wavelength Division Multiple Access Optical Networks, Andrea Borella, Giovanni Cancellieri, and Franco Chiaraluce

For further information on these and other Artech House titles, including previously considered out-of-print books now available through our In-Print-Forever® (IPF®) program, contact:

Artech House
685 Canton Street
Norwood, MA 02062
Phone: 781-769-9750
Fax: 781-769-6334
e-mail: artech@artechhouse.com

Artech House
16 Sussex Street
London SW1V 4RW UK
Phone: +44 (0)20 7596-8750
Fax: +44 (0)20 7630-0166
e-mail: artech-uk@artechhouse.com

Find us on the World Wide Web at: www.artechhouse.com